AF883723

Emergence

A Mathematical Journey from the Big Bang to Sustainable Global Civilization

Pravir Malik, Ph.D.

Copyright © 2017 Pravir Malik
All rights reserved.

ISBN-13: 978-0-9903574-4-5

To Sri Aurobindo

Author's Note

The point of this book is to illustrate a ubiquitous four-fold functionality that exists in the structuring of this universe. What has emerged through the process of time starting with a Big Bang to modern Civilization is not random but is ordered around this four-fold functionality of Presence, Knowledge, Harmony, and Service.

These core functional properties are observed in the nature of reality as set up by light itself. The process of the Big Bang, through the slowing down of light, and the inter-relation between implicit and projected properties of light, created a seed-equation that has subsequently propelled the process of emergence through numerous iterations.

These properties subsequently animate several layers of emergence starting from the 'field' level, through the 'quantum particle' level, the level of 'atoms', 'cells' and even 'civilizations'.

This book will present a mathematics that elaborates the relationship, the four-fold functional relationship to be precise, which exists between one layer of emergence and the next. As such the expression of the universe will be seen as unanimous with the expression of an evolving four-fold functional richness.

The notion of emergence presented here should be coupled with the fundamentally different notion of emergence commonly thought to be true of complex adaptive systems in general. In the latter, emergence is seen as unpredictable in that given the reality of quantum particles, say, it is impossible to predict that a fundamentally different and foundational physical form, that of atoms, would surface. Similarly it is impossible from just observing the dynamics of atoms that something like cells would emerge, or at a later stage even something as complex as civilization.

By coupling the two notions of emergence form can be seen as a 'switch' of sorts, or a receptacle, by which certain functionality can manifest. In this view the properties of Presence, Knowledge, Harmony, and Service take on a different meaning and put us face-to-face with a vast Complex Adaptive System that is our universe.

By examining the mathematics created from the Big Bang seed-equation, this book suggests that there are mathematical equations that need to be fulfilled for a sustainable global civilization to successfully emerge. Further, this book suggests an algorithm, also based on the Big Bang seed-equation, that ties together or rather propels the emergence of each subsequent and more complex layer.

Note that there are details in each of the layers that scientists and specialists – whether those focused on optics, or astrophysics, or quantum physics, or chemistry, or cell biology, or anthropology, or history, amongst other areas - may find incomplete or even misplaced. To these individuals I extend a special invitation to collaborate on working towards a more comprehensive description.

Pravir Malik,
San Francisco

SECTION 1: LIGHT, A BIG BANG & SEED-EQUATION, STRUCTURED TIME & SPACE, AND EMERGENCE

This section lays the foundation for Emergence by examining the nature of light. It is proposed that when light slows down a Big Bang and a Seed-Equation results. But there are properties in Light that structure space and time, and the dynamics of space and time themselves lead to a reality of emergence.

- Light @c
- Light @Infinite
- The Big Bang & The Seed-Equation
- What is Light?
- Structured Time, Structured Space, and the dynamic of Emergence

Chapter 1.1: Light@c

Everything is made from light. But how can this be. How can the dense material objects that make up our world – the stones, cars, buildings – be made from light? How can our thoughts, our emotions, our cells and bodies be made from light? How can all the animals and plants and clouds be made from light? Here is a story that explores how.

To begin this story think for a minute as to how fast light is traveling. If we turn on a light switch in a room it immediately gets filled with light. If we turn on a light torch on a dark night it immediately lights up a path through the darkness. And yet the light from the sun takes about eight minutes to reach the earth. So while light is moving incredibly fast it is still not moving at the infinite speed we may think it is moving at.

Now, it is because the light is not traveling at an infinite speed that the world of matter can appear. All matter is made from atoms. So let us enter into the world of a hypothetical atom in the process of creation to explore how this might be. Imagine that some form of light originates from what will become the nucleus within an atom. That light would take some time, however small, to travel any minute distance within the forming atom. So for that light to express itself, to be noticed, it has to take some amount of a fraction of time. During that fraction of a time there is a build up of energy and it is this build up of energy that forms a packet, as it were, that allows matter to form. It is this build up on energy that variously expresses itself as quarks and leptons and bosons and subsequently atoms– the very building blocks of everything that is.

So, in this view, it is because of the finite speed of light that matter is created. That is not to say that matter will be created wherever light is traveling at a finite speed, but the possibility exists, so long as other conditions are also fulfilled. But let us leave aside the other conditions and the exact process of such a creation for now.

But what else is implicit or made possible because of the finite speed of light? Well, imagine traveling on a ray of light from the sun to the earth. Imagine that you are in minute 4 of the approximately 8 minute journey. As you look back you will see that 4 minutes in the past you were at the sun. 4 minutes in the future you will be at the earth. And in the present moment you are somewhere between the sun and the earth. So this limited speed of light already creates the concept of time and specifically of the past, the present, and the future.

So, four incredibly fundamental things are created because of the finite speed of light: matter, the past, the present, and the future. It could be said, therefore, that this matter-based time-real universe is a result of the finite speed of light.

Chapter 1.2: Light @Infinity

But what if light can be experienced at other speeds? What if light can travel at other finite or an infinite speed beyond what we experience it at, which by the way is an astronomical 186,000 miles per second in a vacuum?

Let us imagine for a minute that light can travel infinitely fast. Think about a big area or volume with a light source at the center. Now since the light travels infinitely fast it will fill up the entire volume instantly. This will be true no matter how large the volume is – it could be the entire solar system, an entire galaxy, or an entire universe. So that light is going to be instantaneously present everywhere – it is going to be omnipresent.

Now, since the light is already present everywhere in whatever volume, that is, since that light has already filled up the entire volume, there is nothing else that can similarly arise there that is not of the nature of light. So the light is all-powerful or omnipotent within that volume.

Further, since the light exists simultaneously everywhere in that space and has a complete knowledge of itself, it therefore has a complete knowledge of that space or of anything that can arise in that space. So it is all-knowing or omniscient in that space.

Finally, it connects everything together instantaneously and holds these connections and the things connected in its nature, so it is all-nurturing or omninurturing within that space.

So, as a result of the infinite speed of light, light appears to have properties of omnipresence, omnipotent, omniscience, and omninurturing.

Chapter 1.3: The Big Bang & The Seed-Equation

We hear about the Big Bang as the start of the universe. In this Big Bang matter is created. But in the examination presented here the creation of matter is nothing other than the result of light traveling at the finite speed, and in this universe at 186,000 miles per second. So we can say that the Big Bang, the apparent start of the universe, is the result of a slowing down of light from an infinite speed to some finite speed.

When light slows down then energy accumulates in packets or quanta and this results in what was inexpressible being able to express itself as matter. This can be thought of in terms of an incredibly rapidly moving stream of water. If the water is traveling so fast then no boundary will be able to contain it and the energy will be continuous over the length of the stream. If it is traveling slower though, then the water will be able to be held by boundaries along the length of the stream. The energy in this case will be discontinuous and will appear in "packets".

In this process of slowing down the implicit omnipresent-omnipotent-omniscient-omninurturing nature of light becomes or transforms intro an explicit or emergent matter-past-present-future nature of light. This process of transformation creates a seed-equation that is present and synonymous with the Big Bang. The seed-equation will be explored in more detail in the section on mathematical foundations.

We have arisen in the field of light that has a finite speed. It is difficult therefore to feel the reality of the omnipresent-omnipotent-omniscient-omninurturing light. We are attuned to perceiving in the material world that has resulted due to the finite speed of light. But if we could step back into the fullness of light in its pure state at infinity from there we would likely see that there can be different universes created as a result of light selectively slowing down to different levels. The slow down to a different level will create a particular kind of universe.

In this view the Big Bang is not an original creative event. It is likely a recurring creative event that allows something innate in light to express itself in time. But what is in light? What is in the omnipresent-omniscient-omnipotent-omninurturing nature of light? Everything we see around us, and an infinite more that we do not see.

Chapter 1.4: What is Light?

When we think of light traveling at the speed of c, there are properties in the reality that emerge as a result of this. These properties as we have examined are of the nature of matter, the past, the present, and the future. These properties are the result of light traveling at c and so can be thought of as emerging from light.

But what do these properties actually mean? And further, how do these properties relate to the apparently quite different properties of omnipresence, omnipotent, omniscience, omninurturing that can appear in a reality where the speed of light is infinite?

Let us consider first the properties of light that emerge when light is traveling at c: the past, the present, the future, and matter.

What is the past? It is the perceivable result of all the work and effort that has taken place so far. It is the foundation upon which the present and future will be built. It represents a status quo, a stability, and even a rigidity, and given that it is the result of the long play of time, it will not easily be persuaded to become another thing. It can be thought of as that which the eye can see when it looks around it. There is a "physicality" to what the eye can see and so the essence of the past is a kind of physical-ness. So, ingrained in light, is this ability to project or create physical-ness.

What is the present? It is the tremendous play of forces of all kinds to express themselves here and now. There is a "vitality" that is present in this play and often it is the most energetic or forceful of the forces that will win out, as opposed to the most insightful or thoughtful. All the tremendous possibility of the future is seeking for expression now and so this essential vitality can also be thought of as a projection or possibility implicit in light.

What is the future? It is the inevitability of what will manifest. The great thoughts, the great ideas, the purpose, the possibilities will sooner or later express themselves in what we call the future. And the essence of this is thoughtfulness or a curiosity or a purpose that we can summarize as an essential "mentality". So embedded in light is this ability to project mentality.

And what is matter? It is the myriad crystallization into apparent diversity of the one essential reality of light, to allow for a play between these different sides or possibilities in an increasingly harmonious interaction. So its essence is "harmony", and this too can be thought of as an essential property projected from or implicit in light.

But what about when light is considered to move at an infinite speed? Then the properties that become apparent, as already explored, are those of omnipresence, omnipotence, omniscience, and omninurturing. But omnipresence has a physical-ness to it, omnipotence has a vitality to it, omniscience a mentality to it, and omninurturing a harmony to it. It may be said that physical-ness comes from omnipresence, vitality comes from omnipotence, mentality comes from omniscience, and matter comes from omninurturing.

So whether light is traveling at an infinite speed or the speed we know as c, there is something about the properties it projects, that in essence is the same. So let us refer to these essential properties made apparent through the worlds that are created, as Presence (from omnipresence), Power (from omnipotence), Knowledge (from omniscience), and Harmony (from omninurturing).

Chapter 1.5: Structured Time, Structured Space, and the Dynamics of Emergence

In considering the worlds that are created due to the way light moves or the play of light, we see properties that are projected because of it. In the finite world, the world that results from light traveling at the speed c, there is, relatively, smallness we can grasp and on which we can build. In the infinite world that results from light traveling at an infinite speed there is a vastness and fullness that is more difficult to grasp.

The notions of time and space are something entirely different in both worlds, and it could be said that space allows the full play of everything meant by Power, Knowledge, Harmony, and Presence to be seeded in it, and that time allows that seeding to flower into fuller forms with its passage.

In other words both space and time are not just abstract concepts but are essentially highly structured to allow physical-ness, vitality, mentality, and matter to become Presence, Power, Knowledge, and Harmony.

Space allows all the possibilities present in Presence, Power, Knowledge, Harmony and seeded in vast diversity, to evolve into more fullness through the time stages of physical-ness, vitality, mentality, and evolving matter.

This then sets up the mechanism or dynamism of emergence: the structure implicit in space and time.

The next section lays the mathematical foundation to begin to explore the notion so emergence more fully.

SECTION 2: LAYING THE MATHEMATICAL FOUNDATION

When the Big Bang happened, light essentially slowed down from an infinite speed to c. In that process a seed-equation was formed that lays out the process and possibilities of emergence through a structuring of space and time.

This section will specifically explore the following foundational mathematics:

- Light Matrix – The Seed-Equation
- Architectural Forces
- Uniqueness of Organizations
- Inherent Dynamics of Any System

Chapter 2.1: Light Matrix – The Seed-Equation

This chapter will explore the Light Matrix, a mathematical expression of the codification in Light, a seed-equation, formed at the time of the Big Bang. The Light Matrix codifies the process of Emergence.

Nature of Light at U

As laid out in Section 1, it is perhaps fair to say that the speed of light has significant implications on the experienced nature of reality. The finiteness, c, at 186,000 miles per second in a vacuum (Perkowitz, 2011), creates an upper bound to the speed with which any object may travel also implying that objective reality will be experienced as a past, a present, a future from the point of view of that object (Einstein, 1995). These characteristics – a past, a present, a future – are implicit in the nature of light and become part of objective reality because of the speed of light. So the way we experience time seems definitely to be determined by c.

Further, c also creates a lower bound when inverted (1/c) being proportional, or arguably even determining Planck's constant, h, that pegs the minimum amount of energy or quanta required for expression at the sub-atomic level. Planck's constant, h, pegging the amount of energy required for expression, therefore may allow matter to form (Lorentz, 1925) and for the reality of nature with a past, present, and future, to also be experienced as a phenomena of connection between seemingly independent islands of matter. This characteristic of 'connection' is therefore also proposed to be implicit in the nature of light and becomes part of objective reality because of the speed of light.

But a 'past' can also be viewed as established reality as defined by what the eye or other lenses of perception can see. It is the perceivable result of all the work and effort that has taken place so far. It is the foundation upon which the present and future will be built. It represents a status quo, a stability, and even a rigidity, and given that it is the result of the long play of time, it will not easily be persuaded to become another thing. It can be thought of as that which the eye can see when it looks around it. There is "physicality" to what the eye can see and so the essence of the past is a kind of physical-ness. So, ingrained in light, is this ability to project or create physical-ness.

The 'present' is the tremendous play of forces of all kinds to express themselves here and now. There is "vitality" that is present in this play and often it is the most energetic or forceful of the forces that will win out, as opposed to the most insightful or thoughtful. All the tremendous possibility of the future is seeking for expression now and so this essential vitality can also be thought of as a projection or possibility implicit in light.

The 'future' is the inevitability of what will manifest. The great thoughts, the great ideas, the purpose, the possibilities will sooner or later express themselves in what we call the future. And the essence of this is thoughtfulness or a curiosity or a purpose that we can summarize as an essential "mentality". So embedded in light is this ability to project mentality.

These implicit characteristics of the nature of light as experienced at the layer of reality so set up by a finite speed of light may hence be summarized by Equation (1), where c_U refers to the speed of light of 186,000 miles per second, that has created the perceived nature of reality, U, as expressed in Equation 2.1.1:

$$c_U: [Physical, Vital, Mental, Connection]$$

Eq. 2.1.1: Implicit Characteristics of the Nature of Light at U

Nature of Light at ∞

Exploring further, it is known however that at quantum levels the nature of reality is characterized by wave-particle duality. Light itself (Feynman, 1985) and matter (De Broglie, 1929) may be experienced as both particles and waves (Ekspong, 2014). But for matter to be experienced as waves implies that 'h' must have become a fraction of itself, $h_{fraction}$, to allow the concentration or possibility of quanta to have dispersed into wave-form. This further implies that c must have become greater than itself, c_N, such that the inequality specified by Equation 2.1.2 holds:

$$c_N > c_U$$

Eq. 2.1.2: Layer N, Layer U Inequality of Speed of Light

Note that what is implied here is that just as there is a nature of reality specified by U that is the result of the speed of light being 186,000 miles per second, so too there is another nature of reality specified by N that is the result of a speed of light greater than 186,000 miles per second.

This is akin to recent developments in physics with the notion of property spaces being separate from but influencing physical space as explored by Nobel Physicist Frank Wilczek (Wilczek, 2016). But further in "Slow Light" Perkowitz's recent treatment of today's breakthroughs in the science of light (Perkowitz, 2011) he states: "Although relativity implies that it's impossible to accelerate an object to the speed of light, the theory may not disallow particles already moving at speed c or greater. In the 1960's, Olexa-Myron P. Bilaniuk of Swarthmore College and E.C. George Sudarshan at Syracuse University began considering how to fit what they called "metaparticles" with speeds greater than c into the relativistic scheme. The approach was extended in 1967 by Gerald Feinberg (Feinberg, 1970), or Rockefeller and Columbia Universities, in his theoretical paper "Possibility of Faster-Than-Light-Particles," Feinberg also introduced the wonderful name "tachyons" for these hypothetical particles, from the Greek word "tachys" meaning swift." Perkowitz goes on to say how a flurry of papers have continued to appear about tachyons.

So light traveling at c_N seems to be possible. Current instrumentation, experience, and normal modes of thinking though, having developed as a bi-product of the characteristics so created in the layer of reality U may be inadequate to access N without appropriate modification. The notion of wave-particle duality already challenges the notion of normal thinking perhaps because wave-like phenomena could be a function of faster than c motion, and particle-like phenomena a function of less than or equal to c motion. That these may be happening simultaneously is reinforced by principles such as complementarity in which experimental observation may allow measurement of one or another but not of both (Whitaker, 2006).

But then taking this trend of a possible increase in the speed of light to its limit, this will result in a speed of light of infinite miles per second. The question is, what is the nature of reality when light is traveling at infinite miles per second? In any space-time continuum light originating at any point will instantaneously have arrived at every other point. Hence light will have a full and immediate *presence* in that space-time continuum. Further, that light will *know* everything that is happening in that space-time completely instantaneously – that is know what is emerging, what is changing, what is diminishing, what may be connected to what, and so on - or have a quality of knowledge. It will connect every object in that space-time completely and therefore have a quality of connection or *harmony*. Finally nothing will be able to resist it or set up a separate reality that excludes it and hence it will have a quality of *power*.

These implicit characteristics of the nature of light as experienced at the layer of reality so set up by an infinite speed of light may hence be summarized by Equation 2.1.3, where c_∞ refers to the speed of light of ∞ miles per second, that has created the perceived nature of reality, ∞:

$$c_\infty: [Presence, Power, Knowledge, Harmony]$$

Eq. 2.1.3: Implicit Characteristics of Nature of Light at ∞ Speed

Transformation from ∞ to U

But by (2.1.3) it can also be noticed that 'physical' is related to presence, 'vital' is related to power, 'mental' is related to knowledge, and 'connection' is related to harmony.

The question then, is how do these apparent qualities at ∞ precipitate or become the physical-vital-mental-connection based diversity experienced at U? This may be achieved through the intervention or action of a couple of mathematical transformations acting on the implicit characteristics of nature of light at ∞ speed as summarized by (2.1.3).

First, the essential characteristics of Presence, Power, Knowledge, Harmony that it is posited exist at every point-instant by virtue of the ubiquity of light at ∞ will need to be expressed as sets with up to infinite elements.

Second, elements in these sets will need to combine together in potentially infinite ways to create a myriad of seeds or signatures that then become the source of the immense diversity experienced at U. This suggests that all that is seen and experienced at U may be nothing other than 'information' or 'content' of light and as such that there are fundamental mathematical symmetries at play where everything at U is essentially the same thing that exists at ∞.

Assuming that the first transformation occurs at a layer of reality K where the speed of light is c_K, such that $c_U < c_K < c_\infty$, this may be expressed by Equation 2.1.4:

$$c_K: [S_{Pr}, S_{Po}, S_K, S_H]$$

Eq. 2.1.4: The First Transformation at Layer K

S_{Pr} signifies 'Set of Presence', S_{Po} signifies 'Set of Power', S_K signifies 'Set of Knowledge', S_H signifies 'Set of Harmony/Nurturing'.

Assuming that the second transformation occurs at a layer of reality N where the speed of light is c_N, such that $c_U < c_N < c_K < c_\infty$, this may be expressed by Equation 2.1.5:

$$c_{N:}\ f(S_{Pr}\ x\ S_{Po}\ x\ S_K\ x\ S_H)$$

Eq. 2.1.5: The Second Transformation at Layer N

The unique seeds are therefore a function, f, of some unique combination of the elements in the four sets S_{Pr}, S_{Po}, S_K, S_H.

The relationship between the layers of light may be modeled by the following matrix in Equation 2.1.6:

$$Light_{Matrix} = \left| \begin{array}{c} c_\infty : [Pr, Po, K, H] \\ (\downarrow R_{C_K} = f(R_{C_\infty})) \\ c_K : [S_{Pr}, S_{Po}, S_K, S_H] \\ (\downarrow R_{C_N} = f(R_{C_K})) \\ c_{N:} f(S_{Pr}\ x\ S_{Po}\ x\ S_K\ x\ S_H) \\ (\downarrow R_{C_U} = f(R_{C_N})) \\ c_U : [P, V, M, C] \end{array} \right|$$

Eq. 2.1.6: Light Matrix

The matrix should be read from the top row down to the bottom row as indicated by the ↓ between rows, and suggests a series of transformations leading from the ubiquitous nature of light implicit in a point – presence, power, knowledge, harmony - to the seeming diversity of matter observed at the layer of reality U which is fundamentally the same presence, power, knowledge, and harmony projected into another form of itself.

The first transformation is summarized by Equation 2.1.7:

$$R_{C_K} = f(R_{C_\infty})$$

Eq. 2.1.7: Light Matrix First Transformation

This is suggesting that the reality at the layer specified by the speed of light c_K, R_{C_K} is a function of the reality at the layer specified by the speed of light c_∞. This transformation translates the essential nature of a point into the sets described in Equation 2.1.4.

The second transformation is summarized by Equation 2.1.8:

$$R_{C_N} = f(R_{C_K})$$

Eq. 2.1.8: Light Matrix Second Transformation

This is suggesting that the reality at the layer specified by the speed of light c_N, R_{C_N} is a function of the reality at the layer specified by the speed of light c_K. This transformation combines elements of the sets into unique seeds as suggested by Equation 2.1.5.

The third transformation is summarized by Equation 2.1.9:

$$R_{C_U} = f(R_{C_N})$$

Eq. 2.1.9: Light Matrix Third Transformation

This is suggesting that the reality at the layer specified by the speed of light c_U, R_{C_U} is a function of the reality at the layer specified by the speed of light c_N. This transformation builds on the unique seeds suggested by (2.1.5) to create the diversity of U as specified by (2.1.1).

In this framework the notion of wave-particle duality hence may become complementary block-field-wave-particle "quadrality" where block refers to phenomenon resident to ∞, field to phenomenon resident to N, wave to phenomenon resident to K, and particle to phenomenon resident to U.

The block is all the reality always present behind the surface and is captured by (2.1.3) or the top line in the Light Matrix as expressed in (2.1.6). The field is captured by (2.1.4) or the second line from the top in (2.1.6) and can be thought of as layers of possibility existing in each of the sets. The wave is captured by the creation of seeds represented by (2.1.5) or by the third line in (2.1.6). The particle that is apparently disconnected from the whole is captured by (2.1.1) or the bottom line in (2.1.6).

So what we arrive at is a fundamental Light Matrix that suggests how Emergence takes place. There is essentially a structure of Space populated by unique seeds. The following chapters get into the details of such emergence.

Chapter 2.2: Architectural Forces

The characteristics embedded in a point as in (2.1.3) suggest a possibility that is hard to fathom. One can only glimpse the extraordinary nature embedded in a point. And yet it can be suggested that this extraordinary nature is barely visible unless the right analytical lens of the sort being suggested here is first set up. Further, it is suggested that this extraordinary nature is responsible for a broader set of architectural forces that exist behind the visible face of things.

Hence, system-presence, system-power, system-knowledge, and system-nurturing that define the nature of every point in our system, become more tangible as a broader set of architectural forces that emanate from each of them.

Considering system-presence, here is a characteristic that appears to be everywhere (Malik, 2009) at the service of all the constructs that develop within it. There is a diligence and perseverance by which any opportunity for progress is seized. Further, if one considers the extraordinary detail that appears in any construct, whether an atom, a body, a planet, or a galaxy, one is struck by the high degree of perfection that surfaces in this presence.

So if one contemplates the nature of this system-presence there is a set of forces that surface. Depicting such a set as $S_{System_{Pr}}$, one can arrive at elements such as Service, Perfection, Diligence, Perseverance, amongst others, that are part of this set. Hence, the set can be described by Equation 2.2.1:

$$S_{System_{Pr}} \ni [Service, Perfection, Diligence, Perseverance, \ldots]$$

Eq 2.2.1: Set of System Presence

Similarly, considering the characteristic of system-power, one can hypothesize that there is a family of forces that emanates from it. The kinds of forces may be thought of as Power, Courage, Adventure, Justice, amongst others. The set for system-power can hence be depicted by Equation 2.2.2:

$$S_{System_P} \ni [Power, Courage, Adventure, Justice, \ldots]$$

Eq 2.2.2: Set of System Power

Similarly, considering the system-knowledge as the root of various powers that emanate from it, one may characterize the set for system-knowledge by Equation 2.2.3:

$$S_{System_K} \ni [Wisdom, Law\ Making, Spread\ of\ Knowledge\ \ldots]$$

Eq 2.2.3: Set of System Knowledge

The set for system-nurturing is depicted by Equation 2.2.4:

$$S_{System_N} \ni [Love, Compassion, Harmony, Relationship\ \ldots]$$

Eq 2.2.4: Set of System Nurturing

Chapter 2.3 describes how unique signatures or seeds for any type of organization can be built by leveraging the elements in the sets of architectural forces.

Chapter 2.3: Uniqueness of Organizations

The hypothesis is that every organization, whether an atom, cell, person, team, corporation, market, or country is unique and that this uniqueness can be specified in terms of elements of the derived sets for power, knowledge, presence, and nurturing. This hypothesis for uniqueness stems from observations at multiple levels.

At the sub-atomic level Nobel Laureate Wolfgang Pauli's 'Pauli Exclusion Principle' states that no two similar fermions, which include fundamental particles with half-integer spin such as protons, neutrons, and electrons, can occupy the same quantum states simultaneously (Pauli, 1964). Spin has to do with the angle that the particle has to rotate through before being symmetrical with its original state. Half-integer spin particles need to rotate through 720 degrees before being symmetrical with their original state. The implication of the Pauli Exclusion Principle is that fundamental structure and consequently stability comes into being at the atomic level, which as is evident in the periodic table allows the separation of function related to form. This stability related to the underlying structure of atoms implies the basis of uniqueness and diversity. In the absence of the Exclusion Principle matter would just be a dense soup (Hawking, 1988) with particles occupying overlapping space.

At the observable level uniqueness is evident from the immense diversity of distinct species on earth (Mora, 2011) estimated to be over 2 million, and further the uniqueness of every member of each species. This member-level uniqueness is suggested by the difference in non-coding regions of the DNA that may vary in their sequence by about 1 to 4 percent, which in turn result in unique protein binding sequences of each human

(Snyder, 2010), as an example, which in turn results in unique observable qualities.

At the astronomical level Einstein's Special Theory of Relativity (Einstein, 1995) suggests that every coordinate system potentially has its own space-time rendering as opposed to there being one absolute space and time. This implies the notion of uniqueness as an implicit property of space.

The four properties explored in the Chapter 2.1 define the source of that innovation. From this source emanate 4 sets of forces that suggest the boundaries of that innovation.

Assuming then that the fount of uniqueness is system-presence, a general equation for organizations that belong to the family of system-presence can be derived. Such uniqueness can be depicted as Sig_x where the subscript 'x' refers to the source family, and 'Sig' or signature to 'uniqueness'. Hence the uniqueness of an organization in the family of system-presence would be notated by $Sig_{SystemPr}$.

In line with the development of properties of a point and the precipitating architectural forces as discussed in Chapter 2.1 and 2.2 respectively, an approach to constructing such uniqueness is to assume a primary factor X that drives the uniqueness that belongs to the set $S_{SystemPr}$. Further, assume that the uniqueness is qualified by a number of secondary factors Y that may belong to any of the 4 sets - $S_{SystemPr}, S_{SystemP}, S_{SystemK}, S_{SystemN}$. The primary factor X would have a greater weightage than any of the secondary factors Y. The weightage of X hence could be depicted by the number 'a', and the weightage of Y a number 'b_{o-n}', such that a > b. Further, the secondary element can repeat from '0 – n' times, and is hence depicted as $\overline{Yb_{0-n}}$.

The equation, Equation 2.3.1, hence for a unique organization derived from the family of system-presence is:

$$Sig_P = Xa + \overline{Yb_{0-n}} \quad where \begin{bmatrix} X \in [S_{System_{Pr}}] \\ Y \in [S_{System_{Pr}}, S_{System_P}, S_{System_K}, S_{System_N}] \\ a, b \; are \; integers; a > b \end{bmatrix}$$

Eq 2.3.1: System Presence Based Unique Organization

Similarly, an equation, Equation 2.3.2, for a unique organization derived from the family of system-power is:

$$Sig_V = Xa + \overline{Yb_{0-n}} \quad where \begin{bmatrix} X \in [S_{System_P}] \\ Y \in [S_{System_{Pr}}, S_{System_P}, S_{System_K}, S_{System_N}] \\ a, b \; are \; integers; a > b \end{bmatrix}$$

Eq 2.3.2: System Power Based Unique Organization

An equation, Equation 2.3.3, for a unique organization derived from the family of system-knowledge is:

$$Sig_M = Xa + \overline{Yb_{0-n}} \quad where \begin{bmatrix} X \in [S_{System_K}] \\ Y \in [S_{System_{Pr}}, S_{System_P}, S_{System_K}, S_{System_N}] \\ a, b \; are \; integers; a > b \end{bmatrix}$$

Eq 2.3.3: System Knowledge Based Unique Organization

An equation, Equation 2.3.4, for a unique organization derived from the family of system-nurturing is:

$$Sig_I = \overline{Xa + Yb_{0-n}} \quad where \begin{bmatrix} X \in [S_{System_N}] \\ Y \in [S_{System_{Pr}}, S_{System_P}, S_{System_K}, S_{System_N}] \\ a, b \; are \; integers; a > b \end{bmatrix}$$

Eq 2.3.4: System Nurturing Based Unique Organization

The four preceding equations can be generalized by Equation 2.3.5:

$$Sig = \overline{Xa + Yb_{0-n}} \quad where \begin{bmatrix} X \in [S_{System_{Pr}}, S_{System_P}, S_{System_K}, S_{System_N}] \\ Y \in [S_{System_{Pr}}, S_{System_P}, S_{System_K}, S_{System_N}] \\ a, b \; are \; integers; a > b \end{bmatrix}$$

Eq 2.3.5: Generalized Equation for Unique Organization

Having considered the structure of uniqueness, the next question is how does such uniqueness emerge? This is discussed in Chapter 2.4.

Chapter 2.4: Emergence of Uniqueness

While the uniqueness of organizations as represented by the Signature is a seed, like any seed there is a process for its emergence (Kaufmann, 1995; Portugali, 2012; Yates, 2012), and the uniqueness will often be hidden or very much behind the scene until certain conditions are fulfilled (Malik, 2009).

The implicit nature of Time and Space suggest a universal developmental model that provides a cue as to the process for emergence (Deep Order Mathematics Videos, 2016). In this model the four sets of architectural forces already described form a pool in space, as it were, from which possibility arises. Possibility itself is unique from point to point and is governed by the Equation for Uniqueness (Equations 2.3.1 through 2.3.5) described in the previous chapter.

Hence it is observed that initially the uniqueness takes a 'physical' form, moving on to a 'vital' form, and then onto a 'mental' form. Once the orientations implicit in each of these phases are assimilated, then the uniqueness takes on an 'integral' form. The integral form is a threshold phase, and allows the uniqueness suggested by the Signature to emerge in fuller force or in its 'force' form. The final phase is the 'contextual form' that allows the signature to act with impunity within a considered context.

Mathematically, if an organization exists at the physical phase, it may be suggested that its signature or uniqueness is modulated by the constant 'π'. π is the seed of a circle or sphere and can be thought of as defining behavior that is tightly bound. Within such a tightly bound volume it will likely not even be apparent what the uniqueness of an organization necessarily is. Assuming the uniqueness to be defined by the derived question *Sig*, the physical-level (P) behavior can be described by the following equation-segment where 'mod' signifies modulated-by:

$$P: Sig * mod\ (\pi)$$

If an organization exists at the vital level, it may be suggested that its uniqueness is modulated by the Euler-constant 'e'. e is at the root of exponential behavior. The vital by definition is about assertive and aggressive growth the symbol of which is 'e'. Hence vital-level (V) modulation (represented by 'mod') can be described by the following equation-segment:

$$V: Sig * mod\ (e)$$

If an organization exists at the mental level, it may be suggested that its uniqueness is modulated by the Gaussian Distribution 'G'. G summarizes rational behavior with a key direction followed by most, and directions more on the edge followed by outliers. Mental-level dynamics are arguably quite similar, and it can be suggested are best modeled by such a distribution (Salkind, 2007). Mental-level (M) modulation (mod) can hence be described by the following equation-segment:

$$M: Sig * mod\ (G)$$

The physical, the vital, and the mental levels are orientations in which patterns of perceiving, being, behaving are set in their ways. Each pattern has its purpose and its limitation and it can be argued that being able to learn from each orientation and yet being able to move beyond that, is the next logical step in any developmental model. The integral level hence, is about being able to leverage each of the patterns that naturally arise at the three preceding levels at will, and about further, being able to integrate these and arrive at new ways of perceiving and being.

Mathematically such behavior may be represented as being an integrative function ($\int x$) where 'x' is the ability to move between the patterns emanating from G, e, π, at will, represented by $\overline{G, e, \pi}$. Integral-level (I) modulation (mod) of uniqueness (Sig) can hence be represented by the following equation-segment:

$$I\text{:}\ \ Sig * mod\ (\int \overline{G, e, \pi}\ )$$

The condition of overcoming any fixed and limiting patterns is the prerequisite for the emergence of 'Force' or for entering into the force-level. At this level the uniqueness behind the particular development being considered can emerge in its purity and become a truly creative dynamic. This aspect of creativity that is in a sense not bound by circumstance may be represented by the constant 'c', the speed of light in a vacuum (Perkowitz, 2011) which is an upper limit of the layer that systems practically operate in. Force-level (F) modulation (mod) of uniqueness (Sig) can hence be represented by the following equation-segment:

$$F\text{:}\ Sig\ * mod\ (c)$$

Once the signature of an organization arises and continues to exercise itself in its purity, it achieves contextual-mastery (C) and is able to exercise itself as though the context it is acting in, that can vary in scale and complexity, were all of the same substance as itself. This equality may be represented by the integrative function '$\int = 1$'. The equation-segment that notates this contextual-level (C) modulation (mod) applied to organizational uniqueness (Sig) is hence:

$$C\text{:}\ Sig * mod\ (\int = 1)$$

Piecing all the equation-segments together the equation for the emergence of uniqueness (Sig_E), where 'X' can be any of the discussed modulations at the respective development-model levels (P, V, M, I, F, C), is hence summarized by Equation 2.4.1:

$$Sig_E = X \begin{vmatrix} C: Sig * mod\ (\int = 1) \\ F: Sig\ mod\ (c) \\ I: Sig\ mod\ (\int \overline{G, e, \pi}\ ) \\ M: Sig * mod\ (G) \\ V: Sig * mod\ (e) \\ P: Sig * mod\ (\pi) \end{vmatrix}$$

Eq 2.4.1: Emergence of Uniqueness

The power of virtual worlds in engaging people and promoting learning as described in "Why Virtual World Matter" (Thomas & Brown, 2009) can be seen as parallel to the emergence of uniqueness as described by Equation 2.4.1. As described by the authors there is a process of selectively leaving behind part of oneself (movement through the P, V, M levels) to recreate one's identity (movement through level I), and then engaging with others in a shared discourse and culture (movement through levels F, C), which becomes very meaningful for people.

Chapter 2.5: Inherent Dynamics of Any System

So far the inherent innovation that exists at the system level and summarized by the nature of a point has been considered. Further, how this deep fount of innovation is present everywhere and how sets that make more practical the range of creative forces available in each of the four components of a point have also been considered. These architectural forces further define the possibility inherent in any system. Leveraging these sets of forces an equation for the uniqueness of an organization, regardless of scale, was arrived at.

In some sense the precipitation of innovation from the barely perceptible nature of the ubiquitous point, to how this reveals a play of forces, to how organizations take their seed and grow from these forces, has been traced.

It is useful to now turn full-circle to return to the initial orientations that allowed so much to be suggested about the nature of innovation in the first place. It is useful to look deeper into the nature of the physical, the vital, the mental, and the integral, and to derive equations that in effect will provide further insight into the dynamics of innovation inherent in these orientations.

Hence, starting with the physical, an equation, Equation 2.5.1, is summarized as:

$$Physical = \begin{bmatrix} M_3 \rightarrow System_{Pr} \\ (\uparrow F \rightarrow I) \\ M_2 \rightarrow S_{Systemi_{Pr}} \\ (\uparrow Sig \rightarrow F) \\ M_1 \rightarrow Sig_P \\ (\uparrow > P_{P)} \\ U \rightarrow Physical_U \end{bmatrix} TC \rightarrow Physical_T$$

$$Where \begin{bmatrix} Physical_U \ni [inertia, lethargy, status\ quo, \ldots] \\ Physical_T \ni [adaptability, durability, strength, \ldots] \end{bmatrix}$$

Eq 2.5.1: Inherent Dynamism in Physical

Essentially this equation is laying out the conditions of moving from the untransformed or negative physical state represented by $Physical_U$ to the transformed or positive physical state represented by $Physical_T$.

The first matrix should be read from the bottom to the top:

$$\begin{bmatrix} M_3 \rightarrow System_{Pr} \\ (\uparrow F \rightarrow I) \\ M_2 \rightarrow S_{System_{Pr}} \\ (\uparrow Sig \rightarrow F) \\ M_1 \rightarrow Sig_P \\ (\uparrow > P_P) \\ U \rightarrow Physical_U \end{bmatrix}$$

Hence, at the bottom is the starting point '$U \rightarrow Physical_U$' which identifies the default or untransformed (U) level of the physical. The next row up, $(\uparrow > P_P)$, states that when the patterns of the untransformed physical (P_P) have been overcome (>), movement to the next level (↑) is facilitated. Breaking through to the next level, $M_1 \rightarrow Sig_P$, allows its dynamics to become active. Hence, the signature or uniqueness of the physical (Sig_P) becomes active at meta-level 1 (M_1). As this signature becomes more like a Force ($Sig \rightarrow F$), the conditions for breakthrough (↑) to the next level are achieved. This next level is referred to as meta-level 2 (M_2), and indicates that the architectural forces represented by the set of system-presence ($S_{System_{Pr}}$) have become more consciously active. When this Force becomes Integral ($F \rightarrow I$) then the conditions for breakthrough (↑) to the next level are achieved. The next level is notated as M_3 for meta-level 3, and the dynamics here indicate that the equation for system-presence becomes active. Becoming active basically means that the respective meta-level dynamic begins to act at the once 'untransformed' level (U) further modifying it. Modification or transformation began when M_1 became active. Transformation is accelerated when M_2 becomes active, and even further accelerated when M_3 becomes active. Note that the matrix is essentially semiotic in that adjacent levels exist in paired relationships and inter-related functionality (Usó-Doménech et al, 2016).

The notion of meta-layers is being explored by contemporary physicists and Erwin Laszlo in his book Self-Actualizing Cosmos (Laszlo, 2014) summarizes some of these developments: "Physicists describe the domain that underlies and embeds the particles, fields, and forces of the universe variously as quantum vacuum, physical spacetime, nuether, zero-point field, grand-unified field, cosmic plenum, or string-net liquid." Note that 'nuether' refers to a sub-quantum level of reality (Pearson, 1997). Laszlo goes on to describe a revolutionary discovery by Nika Arkani-Hamed of Princeton's Institute of Advanced Study, of a geometrical object, the amplituhedron (Arkani-Hamid et al, 2012), which is not in space-time but governs space-time so that it "appears that spatio-temporal phenomena are the consequence of geometrical relationships in a deeper dimension of physical reality". A deeper dimension of a physical layer suggests synonymity with a meta-layer.

The rate of the transformation can be better envisioned when considering action of the Transformation Circle, or TC. The TC can be thought of as 4 concentric circles, with M_3 at the center. M_3 is surrounded by M_2, which is surrounded by M_1. The outer circle is U. If TC is considered to be a clock, than at time 't = 0', the physical' can be thought of as being entirely in U. The clock starts ticking only when some initial patterns P_P are overcome $(> P_P)$. From this point on as time proceeds the conditions for breakthrough become riper, and a sinusoidal wave begins to integrate more of the concentric circles together. The sinusoid wave (sin) is itself modulated by an euler function, e^x, where 'x' is determined by the strength to overcome patterns (↑) which will likely vary over time but will likely tend to be positive once the clock has started ticking because of the joy experienced with progressive movement. Being that the limit is the outer boundary of the concentric circles, there is further modulation by π until the 4 concentric circles have been integrated. TC, hence, may be represented by Equation 2.5.2:

$$TC \equiv (> P_P) \rightarrow \text{mod}(\sin, e^x, \pi)$$

Eq 2.5.2: Transformation Circle

Hence, the initial nature of the physical that may be characterized by the set comprising of elements such as inertia, lethargy, acceptance of the status quo, amongst other such elements ($Physical_U \ni [inertia, lethargy, status\ quo, \ldots]$), transforms into a physical more characterized by elements such as adaptability, durability, strength, and so on ($Physical_T \ni [adaptability, durability, strength, \ldots$). This transformation represents the inherent innovation-dynamic within the Physical.

Similarly, the equation for the 'Vital', Equation 2.5.3, also shows the built-in transformation that represents the innovation-dynamic within the vital:

$$Vital = \begin{bmatrix} M_3 \rightarrow System_P \\ (\uparrow F \rightarrow I) \\ M_2 \rightarrow S_{System_P} \\ (\uparrow Sig \rightarrow F) \\ M_1 \rightarrow Sig_V \\ (\uparrow > P_{V)} \\ U \rightarrow Vital_U \end{bmatrix} TC \rightarrow Vital_T \,,$$

$$Where \begin{bmatrix} Vital_U \ni [aggression, self\ centeredness, exploitation, \ldots] \\ Vital_T \ni [energy, support, adventure, enthusiasm, \ldots] \end{bmatrix}$$

Eq 2.5.3: Inherent Dynamism of Vital

The equation for the 'Mental', Equation 2.5.4, is similarly summarized as:

$$Mental = \begin{bmatrix} M_3 \rightarrow System_S \\ (\uparrow F \rightarrow I) \\ M_2 \rightarrow S_{System_S} \\ (\uparrow Sig \rightarrow F) \\ M_1 \rightarrow Sig_M \\ (\uparrow > P_{M)} \\ U \rightarrow Mental_U \end{bmatrix} TC \rightarrow Mental_T$$

$$Where \begin{bmatrix} Mental_U \ni [fixation, fundamentalism, fragmentation, \ldots] \\ Mental_T \ni [understanding, imagination, inspiration, \ldots] \end{bmatrix}$$

Eq 2.5.4: Inherent Dynamism of Mental

The equation for the 'Integral', Equation 2.5.5, is similarly summarized as:

$$Integral = \begin{bmatrix} M_3 \rightarrow System_N \\ (\uparrow F \rightarrow I) \\ M_2 \rightarrow S_{System_N} \\ (\uparrow Sig \rightarrow F) \\ M_1 \rightarrow Sig_I \\ (\uparrow > P_{I)} \\ U \rightarrow Integral_U \end{bmatrix} TC \rightarrow Integral_T$$

$$Where \begin{bmatrix} Integral_U \ni [possession, usurpation, hidden\ agendas, \ldots] \\ Integral_T \ni [appreciation, shift\ POV, MPV, synthesis, \ldots] \end{bmatrix}$$

Eq 2.5.5: Inherent Dynamism of Integral

The preceding equations can be generalized by Equation 2.5.6:

$$Innovation_{orientation-x} = \begin{bmatrix} M_3 \rightarrow System_X \\ (\uparrow F \rightarrow I) \\ M_2 \rightarrow S_{System_X} \\ (\uparrow Sig \rightarrow F) \\ M_1 \rightarrow Sig_x \\ (\uparrow > P_{x)} \\ U \rightarrow x_U \end{bmatrix} TC \rightarrow x_T\ , where \begin{bmatrix} x_U \ni [\ldots] \\ x_T \ni [\ldots] \end{bmatrix}$$

Eq 2.5.6: Generalized Equation of Innovation

In this generalized equation, $Innovation_{orientation-x}$, refers to the inherent innovation within a specific orientation. Orientation refers to the physical, the vital, the mental, or the integral.

Further, the notion of a core-matrix can be summarized by the following equation, Equation 2.5.7:

$$
Core\_matrix = \begin{bmatrix} M_3 \rightarrow System_X \\ (\uparrow F \rightarrow I) \\ M_2 \rightarrow S_{System_X} \\ (\uparrow Sig \rightarrow F) \\ M_1 \rightarrow Sig_x \\ (\uparrow > P_{x)} \\ U \rightarrow x_U \end{bmatrix}
$$

Eq 2.5.7: Core Matrix

One of the corollaries of the Generalized Equation of Innovation is that if the source of innovation is more influenced by a meta-level there will be simultaneity of innovation or emergence that becomes apparent at U. The higher the meta-level, the more likely that this simultaneity will be wider spread. In his book Where Good Ideas Come From: The Natural History of Innovation (Johnson, 2010) Johnson states: "A brilliant idea occurs to a scientist or inventor somewhere in the world, and he goes public with his remarkable finding, only to discover that three other minds had independently come up with the same idea in the past year." He refers to an essay "Are Inventions Inevitable" (Ogburn & Thomas, 1922) which uncovered 148 instances of similar yet independent innovation, most of them occurring within the same decade. Some examples include sunspots that were simultaneously discovered in 1611 by four scientists in four different countries, and the law of conservation of energy that was formulated separately four times in the late 1840s, amongst numerous other example.

SECTION 3: ELECTROMAGNETIC SPECTRUM

The electromagnetic spectrum is a technical way to refer to light, essentially because amongst its range of properties it displays a tangible and simultaneous electric and magnetic or "electromagnetic" reality as well. So as light becomes more concrete to us or as the possibilities within it begin to emerge, one of the first forms it takes is as the electromagnetic spectrum. It must be the case that the previously surfaced properties of light – Presence, Power, Knowledge, and Harmony – emerge so as to define the very architecture of the electromagnetic spectrum.

Let us look at this in more detail.

Chapter 3.1: Harmony

To begin with, we had already looked at how the speed of light sets up the nature of reality because of its speed. So for light traveling at c, past, present, future, and the notion of separation due to creation of islands of matter is the reality. It seems then that c architects the possibility of interaction in our system and can therefore be thought of a projection from the property of Harmony to create the basis for a matter-based harmony. So being, it can be suggested that the nature of the resultant interactions allows matter-based organizations, regardless of scale, to come into their own, to grow into their boundaries, and to form bonds based on the sense of being separated from other perceived organizations. This notion of forming bonds seems also to be related to the "magnetic" in electromagnetic.

In equation form, as in Equation 3.1.1 it would therefore be possible to specify the nature of reality so set up by the electromagnetic spectrum (EM_Spectrum) moving at c:

$$EM\_Spectrum_{Speed} = \frac{Xa +}{\overline{Yb_{0-n}}} \quad where \begin{bmatrix} X \in [S_{System_N}] \\ Y \in [S_{System_{P_T}}, S_{System_P}, S_{System_K}, S_{System_N}] \\ a, b \text{ are integers}; a > b \end{bmatrix}$$

Eq 3.1.1: Speed of EM Spectrum

The notion of 'connection', 'growing into one's own', 'form bonds', amongst other attributes of such a reality can be seen as elements of the four sets of architectural forces. Hence 3.1.1 suggests the mathematical equation that so defines the nature of reality due to the speed of c of the electromagnetic spectrum.

This also implies that as the speed of c changes, the nature of relationship in such a reality will also change.

Perhaps it is that there are several different relationships, or types of harmonies possible.

Leveraging the previously defined inequality for C_X,

$Inequality_C$: $C_U < C_{M_1} < C_{M_2} < C_{M_3}$

It may be suggested that the nature of the 'layer' will change. This is depicted in Equations 3.1.2 through 3.1.5:

$$[Nature(U)]_{C_U} =$$

$$\frac{Xa +}{Yb_{0-n}} \ where \begin{bmatrix} X \in [S_{System_N}] \\ Y \in [S_{System_{P_r}}, S_{System_P}, S_{System_K}, S_{System_N}] \\ a, b \ are \ integers; a > b \end{bmatrix}$$

Eq 3.1.2: Nature of U as a Function of C(U)

$$[Nature(M_1)]_{C_{M_1}} =$$

$$\frac{Xa +}{Yb_{0-n}} \ where \begin{bmatrix} X \in [S_{System_N}] \\ Y \in [S_{System_{P_r}}, S_{System_P}, S_{System_K}, S_{System_N}] \\ a, b \ are \ integers; a > b \end{bmatrix}$$

Eq 3.1.3: Nature of M1 as a Function of C(M1)

$$[Nature(M_2)]_{C_{M_2}} = Xa + \overline{Yb_{0-n}} \quad where \begin{bmatrix} X \in [S_{System_N}] \\ Y \in [S_{System_{Pr}}, S_{System_P}, S_{System_K}, S_{System_N}] \\ a, b \text{ are integers}; a > b \end{bmatrix}$$

Eq 3.1.4: Nature of M2 as a Function of C(M2)

$$[Nature(M_3)]_{C_{M_3}} = Xa + \overline{Yb_{0-n}} \quad where \begin{bmatrix} X \in [S_{System_N}] \\ Y \in [S_{System_{Pr}}, S_{System_P}, S_{System_K}, S_{System_N}] \\ a, b \text{ are integers}; a > b \end{bmatrix}$$

Eq 3.1.5: Nature of M3 as a Function of C(M3)

In every one of the preceding cases what is to be noticed is that the primary element X derives from the architectural set of nurturing, S_{System_N}.

Chapter 3.2: Knowledge

The electromagnetic spectrum contains a range of waves embedded in it. These waves enable many different applications with practical utility apparent in every day life. So for example there is a region in the electromagnetic spectrum that we call radio waves, and others that we know as microwave, infrared visible light, ultraviolet, x-rays and gamma rays that each make possible many technologies that we use every single day. These ranges essentially code a range of technological-possibility, and therefore the wave-range within the electromagnetic spectrum can be thought of as encoding or of expressing some kind of precipitation or projection or emergence of the property of Knowledge. Or put another way, the property of Knowledge that is found in light, emerges as the wave-range implicit in the electromagnetic spectrum.

The EM spectrum itself, with its vast range of natures from gamma rays through visible light through radio waves, with its implicitness of time-space possibility as suggested by ν and λ, may be thought of as an arrangement of archetypes of what is possible in systems, and therefore is perhaps a precipitation of system-knowledge, as in Equation 3.2.1:

$$EM\_Spectrum_{Structure} = \frac{Xa +}{Yb_{0-n}} \; where \begin{bmatrix} X \in [S_{System_K}] \\ Y \in [S_{System_{Pr}}, S_{System_P}, S_{System_K}, S_{System_N}] \\ a, b \text{ are integers}; a > b \end{bmatrix}$$

Eq 3.2.1: Structure of EM Spectrum

What this also implies is that the significance or intent of the different types of waves that exist can also be expressed by this general equation where the X and Y elements will vary. What precisely these elements are will need to be worked out. A few representative equations, Equations 3.2.2 through 3.2.5 follow:

$$Gamma\ Rays\ _{Intent} =$$

$$\frac{Xa+}{Yb_{0-n}}\ where \begin{bmatrix} X\ \in [S_{System_K}] \\ Y \in [S_{System_{P_r}}, S_{System_P}, S_{System_K}, S_{System_N}] \\ a, b\ are\ integers; a > b \end{bmatrix}$$

Eq 3.2.2: Gamma Rays Intent

Knowing that some of the applications of Gamma Rays are in sterilizing and radiotherapy, these can be attributed as elements to the architectural sets. The totality of the use of can be thought of as the 'intent' of Gamma Rays. An understanding of the uses will allow a full set of the secondary elements to be mapped thus allowing (3.2.2) to be reverse-engineered.

Similarly each of the archetypes present in the electromagnetic spectrum can be mapped out. Sample mapping of intent include x-rays, infrared, and microwaves:

$$X - Rays_{Intent} =$$

$$\frac{Xa+}{Yb_{0-n}}\ where \begin{bmatrix} X\ \in [S_{System_K}] \\ Y \in [S_{System_{P_r}}, S_{System_P}, S_{System_K}, S_{System_N}] \\ a, b\ are\ integers; a > b \end{bmatrix}$$

Eq 3.2.3: X-Rays Intent

$$Infrared_{Intent} =$$

$$\frac{Xa+}{Yb_{0-n}}\ where \begin{bmatrix} X\ \in [S_{System_K}] \\ Y \in [S_{System_{P_r}}, S_{System_P}, S_{System_K}, S_{System_N}] \\ a, b\ are\ integers; a > b \end{bmatrix}$$

Eq 3.2.4: Infrared Intent

$$Microwaves_{\ Intent} = \frac{Xa +}{\overline{Yb_{0-n}}} \ where \begin{bmatrix} X \ \in [S_{System_K}] \\ Y \in [S_{System_{P_T}}, S_{System_P}, S_{System_K}, S_{System_N}] \\ a, b \ are \ integers; a > b \end{bmatrix}$$

Eq 3.2.5: Microwaves Intent

Chapter 3.3: Power

The range of different wave-types or wavelengths implicit in the electromagnetic spectrum also moves with different frequencies. Since the speed of light is a constant, and it is known that speed is the product of frequency and wavelength, the greater the wavelength the lower will be the frequency of the wave-type. Conversely the less the wavelength the higher will be the frequency of the wave-type. And energy or power of a wave-type will depend on its frequency.

Hence, gamma rays that have a lower wavelength will have a higher frequency, and higher energy associated with it. Radio waves on the other hand that have a higher wavelength will have a lower frequency and therefore lower energy associated with it. So implicit in the electromagnetic spectrum is a gradient of energy. But it is also known that the penetration power is dependent on frequency. Therefore, the higher the frequency or energy, the higher will be the power of the wave-type. Another way to say this is that the property of Power in Light emerges as the energy-gradient in the electromagnetic spectrum. This Power aspect, by the way, seems to have been captured by the "electro" in electromagnetic.

The energy-gradient implicit in the EM spectrum suggests the power and energy with which knowledge moves and is perhaps a precipitation of system-power, as in Equation 3.3.1:

$$EM\_Spectrum_{Energy} = \frac{Xa + }{Yb_{0-n}} \quad where \begin{bmatrix} X \in [S_{System_P}] \\ Y \in [S_{System_{Pr}}, S_{System_P}, S_{System_K}, S_{System_N}] \\ a, b \text{ are integers}; a > b \end{bmatrix}$$

Eq 3.3.1: EM Spectrum Energy

This energy as suggested is directly proportional to the frequency, of which there is an infinite range as predicted by Maxwell's equations. Different frequencies have different penetration profiles (HyperPhysics, 2016) and it may be suggested that the nature of the energy is also a precipitation of a range of to be determined meta-functions as suggested in some sample Equations 3.3.2 through 3.3.6. Hence:

$$Gamma\ Rays_{\ Nature\ of\ Energy} = \frac{Xa+}{Yb_{0-n}}\ where \begin{bmatrix} X \in [S_{System_P}] \\ Y \in [S_{System_{P_T}}, S_{System_P}, S_{System_K}, S_{System_N}] \\ a, b\ are\ integers; a > b \end{bmatrix}$$

Eq 3.3.2: Gamma Rays Nature of Energy

$$X - Rays_{Nature\ of\ Energy} = \frac{Xa+}{Yb_{0-n}}\ where \begin{bmatrix} X \in [S_{System_P}] \\ Y \in [S_{System_{P_T}}, S_{System_P}, S_{System_K}, S_{System_N}] \\ a, b\ are\ integers; a > b \end{bmatrix}$$

Eq 3.3.3: X-Rays Nature of Energy

$$Ultraviolet_{\ Nature\ of\ Energy} = \frac{Xa+}{Yb_{0-n}}\ where \begin{bmatrix} X \in [S_{System_P}] \\ Y \in [S_{System_{P_T}}, S_{System_P}, S_{System_K}, S_{System_N}] \\ a, b\ are\ integers; a > b \end{bmatrix}$$

Eq 3.3.4: Ultraviolet Nature of Energy

$$White\ Light_{Nature\ of\ Energy} = \frac{Xa+}{Yb_{0-n}} \quad where \begin{bmatrix} X \in [S_{System_P}] \\ Y \in [S_{System_{P_r}}, S_{System_P}, S_{System_K}, S_{System_N}] \\ a, b\ are\ integers; a > b \end{bmatrix}$$

Eq 3.3.5: White Light Nature of Energy

$$AM\ Radio\ Waves_{Nature\ of\ Energy} = \frac{Xa+}{Yb_{0-n}} \quad where \begin{bmatrix} X \in [S_{System_P}] \\ Y \in [S_{System_{P_r}}, S_{System_P}, S_{System_K}, S_{System_N}] \\ a, b\ are\ integers; a > b \end{bmatrix}$$

Eq 3.3.6: AM Radio Waves Nature of Energy

Chapter 3.4: Presence

If there is a large range of frequencies implicit in the electromagnetic spectrum then there is also the possibility of different types of masses implicit in the electromagnetic spectrum. Frequency determines energy, and mass and energy are related through Einstein's famous MC-squared equation. So pushing a little further it is not just that mass and energy are related, but a different kind of frequency or wave-type potentially perhaps allows a different type of mass to emerge. So the possibility of different types of mass seems to be related to the property of Presence in Light. In other words the property of Presence emerges as the possibility of different types of masses as suggested by the range of mass-possibilities that can emerge from the electromagnetic spectrum.

Mass can be thought of as a container at U within which all possibility happens. In other words it can be thought of as a precipitation or emergence of system-presence as depicted in Equation 3.4.1:

$$EM\_Spectrum_{Mass\_Possibility} = \frac{Xa +}{Yb_{0-n}} \quad where \begin{bmatrix} X \in [S_{System_{Pr}}] \\ Y \in [S_{System_{Pr}}, S_{System_{P}}, S_{System_{K}}, S_{System_{N}}] \\ a, b \; are \; integers; a > b \end{bmatrix}$$

Eq 3.4.1: EM Spectrum Mass Possibility

If the frequencies are infinite, then the possibility of the 'types' of masses or matter is also infinite. The mystery of 'Dark Matter' suggested by scientists to be 27% of our universe, as opposed to 5% of visible matter (NASA-darkmatter, 2016) may have some relation to this.

Hypothetically the Equations 3.4.2 through 3.4.6 depict a range of mass possibilities:

$$Gamma\ Rays_{\ Mass\_Possibility} =$$

$$\frac{Xa +}{Yb_{0-n}} \quad where \begin{bmatrix} X \in [S_{System_{Pr}}] \\ Y \in [S_{System_{Pr}}, S_{System_P}, S_{System_K}, S_{System_N}] \\ a, b \ are \ integers; a > b \end{bmatrix}$$

Eq 3.4.2: Gamma Rays Mass Possibility

$$Ultraviolet_{\ Mass\_Possibility} =$$

$$\frac{Xa +}{Yb_{0-n}} \quad where \begin{bmatrix} X \in [S_{System_{Pr}}] \\ Y \in [S_{System_{Pr}}, S_{System_P}, S_{System_K}, S_{System_N}] \\ a, b \ are \ integers; a > b \end{bmatrix}$$

Eq 3.4.3: Ultraviolet Mass Possibility

$$Blue\ Light_{\ Mass\_Possibility} =$$

$$\frac{Xa +}{Yb_{0-n}} \quad where \begin{bmatrix} X \in [S_{System_{Pr}}] \\ Y \in [S_{System_{Pr}}, S_{System_P}, S_{System_K}, S_{System_N}] \\ a, b \ are \ integers; a > b \end{bmatrix}$$

Eq 3.4.4: Blue Light Mass Possibility

$$Microwaves_{\ Mass\_Possibility} =$$

$$\frac{Xa +}{Yb_{0-n}} \quad where \begin{bmatrix} X \in [S_{System_{Pr}}] \\ Y \in [S_{System_{Pr}}, S_{System_P}, S_{System_K}, S_{System_N}] \\ a, b \ are \ integers; a > b \end{bmatrix}$$

Eq 3.4.5: Microwaves Mass Possibility

$$FM\ Radio\ Waves_{Mass\_Possibility} = \frac{Xa +}{Yb_{0-n}} \quad where \begin{bmatrix} X \in [S_{System_{Pr}}] \\ Y \in [S_{System_{Pr}}, S_{System_P}, S_{System_K}, S_{System_N}] \\ a, b\ are\ integers; a > b \end{bmatrix}$$

Eq 3.4.6: FM Radio Waves Mass Possibility

SECTION 4: EMERGENCE OF MATTER

So we find that the four underlying properties of Light that we call Harmony, Knowledge, Power, and Presence are of the essence of the speed with which the electromagnetic spectrum moves, the wave-range within the electromagnetic spectrum, the energy-gradient within the electromagnetic spectrum, and the mass-possibilities due to the electromagnetic spectrum, respectively. Further the description – electromagnetic – seems to have captured the Power-Harmony aspects implicit in light. In reality the electro-magnetic spectrum can likely be more completely described as wavearchetype-electro-magnetic-masspotential spectrum.

But further, we will find that layers of matter – quantum particles, which include bosons, and atoms – are also structured or emerge along the same property-lines or property-families of light.

Chapter 4.1: Quantum particles

The electromagnetic spectrum or perhaps more accurately, the wavearchetype-electro-magnetic-masspotential spectrum, describes a first layer or translation of light into manifestation. This spectrum or field is an example of what stands behind visible matter and perhaps it is fair to say that all matter emerges from such spectrums.

As already explored the finite speed of such spectrums allows the build of energies or quanta, and this in turn may express itself as a series or as an array of quantum particles. The vast number of quantum particles that have so far been discovered has in turn yielded what is known as the Standard Model. This model is made up of what is called Quarks, Leptons, and Bosons, and a Higgs-Boson (Cottingham, 2007).

But if we look at these four fundamental quantum categories of particles from a property-based, or "functional" viewpoint a different chapter in the story of light emerges.

Quarks

For instance, it can be seen first that the nucleus of an atom is made up of a combination of quarks. Specifically, a proton is composed of two "up" quarks and one "down" quark. Quarks have unusual names – up, down, charm, strange, top, bottom, with each subsequent pair belonging to a different "generation" of quarks. A neutron is composed of two "down" quarks and one "up" quark. Protons and neutrons together make up the nucleus of an atom. But we also know that the number of protons in the nucleus specifies the Atomic Number of an atom. Atomic number in turn uniquely identifies the element from the periodic table. Hence, an atomic number of 47, for example, specifies that the element is Silver. In other words it can be suggested that the unique properties of an element, the knowledge of what it is and how it will behave in the universe, is related to the quark. It may be suggested that quarks, therefore, are associated with the precipitation or emergence of Light's property of Knowledge in the quantum world.

Hence, it could be that the signature for the family of quarks, as in Equation 4.1.1, is:

$$Sig_{quarks} = Xa + \overline{Yb_{0-n}} \quad where \begin{bmatrix} X \in [S_{System_K}] \\ Y \in [S_{System_{Pr}}, S_{System_P}, S_{System_K}, S_{System_N}] \\ a, b \ are \ integers; a > b \end{bmatrix}$$

Eq 4.1.1: Generalized Signature of Quarks

As can be seen the primary element X is derived from the set of knowledge, S_{System_K}. Various elements, derived from the four sets would define the behavior of quarks and could be functions such as 'composite-arrangements', 'specifying attributes' amongst others, hence collectively describing quarks' way of being.

Leptons

When considering leptons it is useful to know that unlike quarks that only exist in composite arrangements with other quarks, leptons are solitary, point-like particles without internal structure (Arabatzis, 2006). The best-known lepton is the electron. So the electron may be considered as a surrogate for the lepton class. The electron appears to be associated with the flow of energy and power. Further they appear to be the adventurers easily leaving the atom they are a part of. They also form locks or bonds with other atoms through the force of attraction and repulsion. In some sense they seem to be a representation or precipitation or emergence of Light's property of Power.

The signature for the family of leptons, as in Equation 4.1.2, is:

$$Sig_{leptons} = Xa + \overline{Yb_{0-n}} \quad where \begin{bmatrix} X \in [S_{System_P}] \\ Y \in [S_{System_{P_T}}, S_{System_P}, S_{System_K}, S_{System_N}] \\ a, b \ are \ integers; a > b \end{bmatrix}$$

Eq 4.1.2: Generalized Signature of Leptons

As can be seen the primary element X is derived from the set of power, S_{System_P}. Various elements, derived from the four sets would define the behavior of leptons and could be functions such as 'adventurer', 'solitary', 'create attraction', amongst others, hence collectively describing leptons' way of being.

Bosons

The bosons on the other hand are thought of as force-carriers. They are what allow all known matter particles to interact. The three fundamental bosons in this category are the photon, the W and Z bosons, and the gluon. The carrier particle of the electromagnetic force or spectrum is the photon. The carrier particle of the strong nuclear force that holds quarks together is the gluon. The carrier particle for the weak interactions, responsible for the decay of massive quarks and leptons into lighter quarks and leptons, are the W and Z bosons.

Bosons can be thought of as the precipitation of what creates relationship and harmony at the quantum level. Hence they can be thought of as the precipitation or emergence of Light's property of Harmony at the quantum level.

The signature for the family of gauge bosons, as in Equation 4.1.3, is:

$$Sig_{bosons} = Xa + \overline{Yb_{0-n}} \quad where \begin{bmatrix} X \in [S_{System_N}] \\ Y \in [S_{System_{P_T}}, S_{System_P}, S_{System_K}, S_{System_N}] \\ a, b \ are \ integers; a > b \end{bmatrix}$$

Eq 4.1.3: Generalized Signature of Bosons

As can be seen the primary element X is derived from the set of nurturing, S_{System_N}. Various elements, derived from the four sets would define the behavior of bosons and could be functions such as 'creating interaction', 'holding together', amongst others, hence collectively describing bosons' way of being.

Equations in this form give a clue as to how to think about the fundamental particles. While the default approach is always to think about physical qualities, equations in the preceding format allow us to think about detailed functions or qualities associated with particles and give insight into how the process of emergence takes place.

Higgs-Boson

This leaves the other discovered fundamental particle the Higgs-Boson. In ordinary matter, most of the mass is contained in atoms, and the majority of the mass of an atom resides in the nucleus, made of protons and neutrons. Protons and neutrons are each made of three quarks. But it is the quarks that get their mass by interacting with what is known as the Higgs field (Olive, 2014). Hence the Higgs-Boson can be thought of as the mass-giver. In other words it is what gives presence to the quarks and it can be thought of as the precipitation or emergence of Light's property of Presence at the quantum level. Just as there are multiple particles in each of the other 'families' it is likely that there will be multiple particles in the Higgs-Boson family. Recent research at CERN indicates that the Higgs-Boson may have a cousin.

The signature for the Higgs-boson and any other similar particle, as in Equation 4.1.4, is:

$$Sig_{Higgs-boson} = Xa + \overline{Yb_{0-n}} \quad where \begin{bmatrix} X \in [S_{System_{Pr}}] \\ Y \in [S_{System_{Pr}}, S_{System_P}, S_{System_K}, S_{System_N}] \\ a, b\ are\ integers; a > b \end{bmatrix}$$

Eq 4.1.4: Generalized Signature of Higgs-Boson

As can be seen the primary element X is derived from the set of presence, $S_{System_{Pr}}$. Various elements, derived from the four sets would define the behavior of Higgs-bosons and could be a function such as 'creating mass', amongst others, hence collectively describing Higgs-bosons' way of being.

Combining all the preceding particle equations it is possible to create a generalized particle equation, as in Equation 4.1.5:

$$Sig_{particle} = Xa + \overline{Yb_{0-n}} \quad where \begin{bmatrix} X \in [S_{System_{Pr}}, S_{System_P}, S_{System_K}, S_{System_N}] \\ Y \in [S_{System_{Pr}}, S_{System_P}, S_{System_K}, S_{System_N}] \\ a, b \text{ are integers}; a > b \end{bmatrix}$$

Eq 4.1.5: Generalized Signature of Particle

So we see that again the underlying properties of light - Knowledge, Power, Harmony, and Presence - emerge as quarks, leptons, bosons, and Higgs-bosons respectively.

Chapter 4.2: Bosons

The bosons as mentioned can be thought of as force-carriers and allow all known matter particles to interact.

But when we look at bosons in more detail there are three fundamental bosons – the photon, the W and Z bosons, the gluon - and one hypothetical boson – the graviton.

Photon

The photon is the carrier particle of the electromagnetic force. But in the scheme of things the electromagnetic force pervades everything and as explored in the section on the electromagnetic spectrum appears to be foundational to this reality we are in. So it could be said that it is related to Presence or is an emergence of Light's property of Presence at the level of quantum force-carriers.

Hence, an equation for the photon, Equation 4.2.1 could be:

$$Sig_{photon} = Xa + \overline{Yb_{0-n}} \quad where \begin{bmatrix} X \in [S_{System_N}] \\ Y \in [S_{System_{P_r}}, S_{System_P}, S_{System_K}, S_{System_N}] \\ a, b \text{ are integers}; a > b \end{bmatrix}$$

Eq 4.2.1: Signature of Photon

While the primary element X, would continue to be the same as X for bosons, as in Equation 4.1.3, there will be additional secondary elements Y that will further qualify the attributes or functionality of photons. As an example such elements could be 'pervasiveness', 'multiple wave handler', amongst others.

Gluon

The gluon is the carrier particle of what is known as the strong nuclear force and holds quarks together in their inherently composite arrangements. But we know that quarks are related to or are an emergence of Knowledge at the quantum-particle level. Hence it must be that the gluon is an emergence of Light's property of Knowledge at the level of quantum force-carriers.

Hence, Equation 4.2.2 is an equation for the gluon:

$$Sig_{gluon} = Xa + \overline{Yb_{0-n}} \ where \begin{bmatrix} X \in [S_{System_N}] \\ Y \in [S_{System_{Pr}}, S_{System_P}, S_{System_K}, S_{System_N}] \\ a, b \ are \ integers; a > b \end{bmatrix}$$

Eq 4.2.2: Signature of Gluon

The primary element X would continue to be the same as in the equation for the boson. There are likely additional secondary elements Y such as 'concentrated', 'intense connection', amongst others that would collectively determine the behavior of gluons.

W and Z Bosons

The W and Z bosons are the carrier particle for the weak interactions, responsible for the decay of massive quarks and leptons into lighter quarks and leptons. But this usually is accompanied by the release of energy and power and so it must be that W and Z bosons are an emergence of Light's property of Power at the level of quantum force-carriers.

Hence, Equation 4.2.3, for W and Z bosons is as follows:

$$Sig_{W,Z\ bosons} = Xa + \overline{Yb_{0-n}} \ where \begin{bmatrix} X \in [S_{System_N}] \\ Y \in [S_{System_{Pr}}, S_{System_P}, S_{System_K}, S_{System_N}] \\ a, b \ are \ integers; a > b \end{bmatrix}$$

Eq 4.2.2: Signature of W and Z Bosons

The primary element X would continue to be the same as in the equation for the boson. There are likely additional secondary elements Y such as 'weak link', 'release of energy', amongst others that would collectively determine the behavior of W and Z bosons.

Graviton

This leaves the hypothetical graviton that is thought to be the carrier particle for the force of gravity. But gravity is what holds astronomical objects together in relationship. Hence it must be that gravitons are an emergence of Light's property of Harmony at the level of quantum force-carriers.

Hence, Equation 4.2.4, for the hypothetical graviton is as follows:

$$Sig_{graviton} = Xa + \overline{Yb_{0-n}} \quad where \begin{bmatrix} X \in [S_{System_N}] \\ Y \in [S_{System_{P_T}}, S_{System_P}, S_{System_K}, S_{System_N}] \\ a, b \; are \; integers; a > b \end{bmatrix}$$

Eq 4.2.2: Signature of W and Z Bosons

The primary element X would continue to be the same as in the equation for the boson. There are likely additional secondary elements Y such as 'mass curving space', 'space defining object movement', amongst others that would collectively determine the behavior of gravitons.

So we see that the underlying properties of light - Presence, Knowledge, Power, and Harmony - emerge as photons, gluons, W and Z bosons, and the hypothetical graviton at the quantum force-carrier level.

Chapter 4.3: Atoms

Quantum particles in turn create atoms, and all known atoms can be classified into one of four groups: The s-group, p-group, d-group, and f-group. But what are these groups and are they too emergences of Light as it continues its journey of crystalizing the possibility or potentiality within it?

s-Group

The s-group consists primarily of what are known as alkali metals and alkali earth metals. These alkali metals are known to easily lose electrons and form what is known as positive ions. When they lose electrons energy is gained, but when the electrons are taken up by other atoms in proximity there is a lot of energy released. Some have referred to these groups as "violent worlds" (Tweed, 2003), and it has been pointed out that stars shine because they are transmuting vast amount of hydrogen into helium, both of which are s-group elements. So one gets the sense that the s-group may be an emergence of Light's property of Power.

Stars and suns are also known to be the crucibles where all the different kinds of atoms are created. So these furnaces of power by virtue of their heat and high pressure are able to force electrons and protons and neutrons to come together to create all the different types of atoms known in the universe.

But philosophically what are the s-group atoms or elements? These are atoms where there is an equal likelihood of an electron being anywhere in a symmetrical sphere around the nucleus. The other groups are all similarly defined by likelihoods of electrons being within a possible pattern around the nucleus. The patterns that distinguishes s-group atoms is a sphere, and since all other patterns can be thought of as occurring within the sphere, in some sense this is like an imprint or precipitation or emergence that allows other kinds of emergences to surface within it. So the elements that are part of the s-group may be thought of as the adventurers with courage who venture into a brave new world to create some foundation by which all other element-creations can follow.

The fact that hydrogen and helium are known to constitute 98% of the Universe (Heiserman, 1991) relative to other elements therefore makes sense in this view, especially since hydrogen and helium provide the fuel with which the star-furnaces manufacture all other elements.

So s-group elements seem to embody functions such as power, energy, adventure, courage, and can be thought of as an emergence of the property of Light to do with Power.

Hence, a series of equations linked to S_{System_P} as the prime set can be suggested, starting with the s-group mapping, as in Equation 4.3.1:

$$Element_{s-group} = \frac{Xa+}{Yb_{0-n}} \quad where \begin{bmatrix} X \in [S_{System_P}] \\ Y \in [S_{System_{P_r}}, S_{System_P}, S_{System_K}, S_{System_N}] \\ a, b \text{ are integers}; a > b \end{bmatrix}$$

Eq 4.3.1: s-Group Element

Further, the equivalent mapping between traditional element groupings and S_{System_P} as in Equations 4.3.2 and 4.3.3 can be specified:

$$Element_{Alkali\ metal} = \frac{Xa+}{Yb_{0-n}} \quad where \begin{bmatrix} X \in [S_{System_P}] \\ Y \in [S_{System_{P_r}}, S_{System_P}, S_{System_K}, S_{System_N}] \\ a, b \text{ are integers}; a > b \end{bmatrix}$$

Eq 4.3.2: Alkali Metal Element

$$Element_{\,Alkali\ earth\ metal} = \frac{Xa +}{Yb_{0-n}} \ where \begin{bmatrix} X \in [S_{System_P}] \\ Y \in [S_{System_{Pr}}, S_{System_P}, S_{System_K}, S_{System_N}] \\ a, b\ are\ integers; a > b \end{bmatrix}$$

Eq 4.3.3: Alkali Earth Metal Element

Further, as a representative element belonging to the Alkali Metal group the equation for Lithium (Li), as in Equation 4.3.4, would be:

$$Element_{\,Lithium} = \frac{Xa +}{Yb_{0-n}} \ where \begin{bmatrix} X \in [S_{System_P}] \\ Y \in [S_{System_{Pr}}, S_{System_P}, S_{System_K}, S_{System_N}] \\ a, b\ are\ integers; a > b \end{bmatrix}$$

Eq 4.3.4: Lithium

Note that the primary element X in all these cases may be a function or attribute along the lines of 'expresses power'. The secondary elements Y in all these would have some elements being the same, and then would have specific differences as one gets into sub-classes and the individual elements themselves.

p-Group

Atoms or elements belonging to the p-group are those with the likelihood of electrons occurring equally on either side of the nucleus, like a dumbbell of sorts.

There are some very significant elements in this group that are part of the metal, metalloid, non-metal, halogen, and noble gas sub-groupings. Carbon, Nitrogen,

Oxygen, and Silicon are some of the sample elements. Looking at the types of elements present in this group it is as though all the element possibilities have been represented within it. It is perhaps that the possibility of ideas behind all elements has emerged in this group and one can hypothesize that this group may be a reflection of the property of Knowledge, forming archetypes from which all other elements are created.

Philosophically, the one spherical shell or probability cloud (s) becoming two smaller shells like a dumbbell (p) signifies the creation of an essential polarity within a unit space. So in a sense the one becoming two is the first instance of variability in space. The two, spaced in 3-dimensions around the nucleus in a sense create six switches becoming an attractor or allowing a vaster number of different kinds of elements to surface. So there is a sense of the 'idea' of the element that comes into focus.

But further, the essential elements that allow both thinking and virtual thinking machines to come into being, are also contained within this group. Carbon is the basis of DNA and of all life. The fact the Silicon, directly below it in the periodic table and therefore sharing essential qualities, is considered the basis of all virtual thinking machines is therefore perhaps significant and may reinforce the notion that the p-group is a precipitation of Light's property of Knowledge.

Hence, a series of equations linked to S_{System_K} as the prime set can be suggested, starting with the p-group mapping, as in Equation 4.3.5:

$$Element_{p-group} =$$

$$Xa + \overline{Yb_{0-n}} \quad where \begin{bmatrix} X \in [S_{System_K}] \\ Y \in [S_{System_{Pr}}, S_{System_P}, S_{System_K}, S_{System_N}] \\ a, b \text{ are integers}; a > b \end{bmatrix}$$

Eq 4.3.5: p-group Element

Further, the equivalent mapping between traditional element groupings and S_{System_K} as in Equations 4.3.6 through 4.3.11 can also be specified:

$$Element_{\ Metal} = \overline{\begin{matrix} Xa + \\ Yb_{0-n} \end{matrix}} \ where \begin{bmatrix} X \in [S_{System_K}] \\ Y \in [S_{System_{Pr}}, S_{System_P}, S_{System_K}, S_{System_N}] \\ a, b\ are\ integers; a > b \end{bmatrix}$$

Eq 4.3.6: Metal Element

$$Element_{\ Metalloid} = \overline{\begin{matrix} Xa + \\ Yb_{0-n} \end{matrix}} \ where \begin{bmatrix} X \in [S_{System_K}] \\ Y \in [S_{System_{Pr}}, S_{System_P}, S_{System_K}, S_{System_N}] \\ a, b\ are\ integers; a > b \end{bmatrix}$$

Eq 4.3.7: Metalloid Element

$$Element_{\ Non-Metal} = \overline{\begin{matrix} Xa + \\ Yb_{0-n} \end{matrix}} \ where \begin{bmatrix} X \in [S_{System_K}] \\ Y \in [S_{System_{Pr}}, S_{System_P}, S_{System_K}, S_{System_N}] \\ a, b\ are\ integers; a > b \end{bmatrix}$$

Eq 4.3.8: Non-Metal Element

$$Element_{Halogen} = \frac{Xa +}{Yb_{0-n}} \text{ where } \begin{bmatrix} X \in [S_{System_K}] \\ Y \in [S_{System_{Pr}}, S_{System_P}, S_{System_K}, S_{System_N}] \\ a, b \text{ are integers}; a > b \end{bmatrix}$$

Eq 4.3.9: Halogen Element

$$Element_{Noble\ Gas} = \frac{Xa +}{Yb_{0-n}} \text{ where } \begin{bmatrix} X \in [S_{System_K}] \\ Y \in [S_{System_{Pr}}, S_{System_P}, S_{System_K}, S_{System_N}] \\ a, b \text{ are integers}; a > b \end{bmatrix}$$

Eq 4.3.10: Noble Gas Element

Further, as a representative element belonging to the Non-Metal group the equation for Carbon (C), as in Equation 4.3.11, would be:

$$Element_{Carbon} = \frac{Xa +}{Yb_{0-n}} \text{ where } \begin{bmatrix} X \in [S_{System_K}] \\ Y \in [S_{System_{Pr}}, S_{System_P}, S_{System_K}, S_{System_N}] \\ a, b \text{ are integers}; a > b \end{bmatrix}$$

Eq 4.3.11: Carbon

Note that the primary element X in all these cases may be a function or attribute along the lines of 'knowledge'. The secondary elements Y in all these would have some elements being the same, and then would have specific differences as one gets into sub-classes and the individual elements themselves.

d-Group

The d-group comprises the Transition Metals. These metals are generally hard and strong, exhibit corrosive resistance, and can be thought of as workhorse elements. Many industrial and well-known elements sit in this group: Titanium, Chromium, Manganese, Iron, Cobalt, Nickel, Copper, Zinc, Silver, Platinum, and Gold, amongst others.

The elements and atoms in the d-group are equally likely to show up in four possible lobes or probability-spaces around the nucleus. Four lobes occurring in multiple possible planes around the nucleus will likely create a space of stability, since there is a possibility of four lobes creating the four vertices of a tetrahedron (Fuller, 1982) that has been positioned as one of the most stable shapes in the universe.

Much of the constructed world around us is created from these elements. Further, most of the series in the group easily lose one or more electrons thereby easily combining with other atoms to form a vast array of compounds. Also, looking more broadly at the function of these elements, it can be seen that these metals exist for service, to help bring about perfection in the constructed world, to help much of the machinery in which they are used, and to assist the processes dependent on them to be completed with diligence. Hence, these transition metals appear to be an emergence of Light's property of Presence.

Therefore, a series of equations linked to $S_{System_{Pr}}$ as the prime set can be suggested, starting with the d-group mapping, as in Equation 4.3.12:

$$Element_{\,d-group} = \frac{Xa + }{Yb_{0-n}} \quad where \begin{bmatrix} X \in [S_{System_{Pr}}] \\ Y \in [S_{System_{Pr}}, S_{System_{P}}, S_{System_{K}}, S_{System_{N}}] \\ a, b \; are \; integers; a > b \end{bmatrix}$$

Eq 4.3.12: d-group Element

Further, the equivalent mapping between traditional element groupings and $S_{System_{Pr}}$, as in Equation 4.3.13, can also be specified:

$$Element_{\ Transition\ Metal} = \frac{Xa +}{Yb_{0-n}} \quad where \begin{bmatrix} X \in [S_{System_{Pr}}] \\ Y \in [S_{System_{Pr}}, S_{System_P}, S_{System_K}, S_{System_N}] \\ a, b\ are\ integers; a > b \end{bmatrix}$$

Eq 4.3.13: Transition Metal Element

Further, as a representative element belonging to the Transition-Metal group the equation for Gold (Au), as in Equation 4.3.14, would be:

$$Element_{\ Gold} = \frac{Xa +}{Yb_{0-n}} \quad where \begin{bmatrix} X \in [S_{System_{Pr}}] \\ Y \in [S_{System_{Pr}}, S_{System_P}, S_{System_K}, S_{System_N}] \\ a, b\ are\ integers; a > b \end{bmatrix}$$

Eq 4.3.14: Gold

Note that the primary element X in all these cases may be a function or attribute along the lines of 'work horse'. The secondary elements Y in all these would have some elements being the same, and then would have specific differences as one gets into sub-classes and the individual elements themselves.

f-Group

The f-group comprises of the Lanthanides and Actinides. Philosophically, elements in the F-group consist of 6 lobes around the nucleus within which an electron may be found. 6 lobes will exist in multiple planes around the nucleus and suggests the notion of extended relationship and collectivity: the attempt to build larger and larger bonds within a small space. Considering this it is likely that the f-group is an emergence of Light's property of Harmony.

Thinking about Lanthanides, some interesting facts may reinforce this notion. First, the spin of electrons in a lanthanides' outer shell is aligned, creating a strong magnetic field. The notion of creating a strong magnetic field seems to be consistent with the notion of engendering a collectivity through the ordered attraction and repulsion of elements. Second, these elements curiously occur together in nature often in the same ores (Gray, 2009) and are chemically interchangeable also suggesting the notion of forming a tight intra-group collectivity.

Actinides on the other hand are inherently radioactive. This implies that these elements have inherently crossed a threshold of stability and have the urge, over their own half-lives, to decompose into other elements. This natural urge may suggest some boundary conditions on the notion of collectivity and nurturing, giving additional insight into the nature of collectivity and nurturing. Further, the entire actinide group, as opposed to the lanthanide group that is inherently stable, is unstable. It is curious that both these should be part of the f-group, and they must provide insight into boundary conditions into the notion of collectivity in elements.

Hence, a series of equations linked to S_{System_N} as the prime set can be suggested, starting with the f-orbital mapping, as in Equation 4.3.15:

$$Element_{f-group} = \overline{Xa + Yb_{0-n}} \quad where \begin{bmatrix} X \in [S_{System_N}] \\ Y \in [S_{System_{Pr}}, S_{System_P}, S_{System_K}, S_{System_N}] \\ a, b \text{ are integers}; a > b \end{bmatrix}$$

Eq 4.3.15: f-group Element

Further, the equivalent mapping between traditional element groupings and S_{System_N}, as in Equations 4.3.16 and 4.3.17, can also be specified:

$$Element_{Lanthanide} = \overline{Xa + Yb_{0-n}} \quad where \begin{bmatrix} X \in [S_{System_N}] \\ Y \in [S_{System_{Pr}}, S_{System_P}, S_{System_K}, S_{System_N}] \\ a, b \text{ are integers}; a > b \end{bmatrix}$$

Eq 4.3.16: Lanthanide Element

$$Element_{Actinide} = \overline{Xa + Yb_{0-n}} \quad where \begin{bmatrix} X \in [S_{System_N}] \\ Y \in [S_{System_{Pr}}, S_{System_P}, S_{System_K}, S_{System_N}] \\ a, b \text{ are integers}; a > b \end{bmatrix}$$

Eq 4.3.17: Actinide Element

Further, as a representative element belonging to the Lanthanide group the equation for Lanthanum (La), as in Equation 4.3.18, would be:

$$Element_{Lanthanum} = \overline{\frac{Xa +}{Yb_{0-n}}} \ where \begin{bmatrix} X \in [S_{System_N}] \\ Y \in [S_{System_{Pr}}, S_{System_P}, S_{System_K}, S_{System_N}] \\ a, b \ are \ integers; a > b \end{bmatrix}$$

Eq 4.3.18: Lanthanum

Note that the primary element X in all these cases may be a function or attribute along the lines of 'experiments in collectivity'. The secondary elements Y in all these would have some elements being the same, and then would have specific differences as one gets into sub-classes and the individual elements themselves.

SECTION 5: EMERGENCE OF LIFE

This section explores the emergence of molecular plans in living cells, and further some fundamental aspects by which we know we are, and would even define being alive – sensations, desires, feelings, and thoughts. Each of these emergences can be seen to architected by the underlying four-fold characteristics of Light.

Chapter 5.1: Cells

In its journey to create a universe light has projected aspects of itself to structure the layers of the electromagnetic field, quantum particles, and atoms. In so doing it has allowed all of matter, and all arrangement of molecules and so even the basis of life to emerge.

But let us see how the properties of light structure the living cell.

This examination will begin by looking at the molecular structure at the cellular level. In 'The Machinery of Life', Goodsell, an Associate Professor of Molecular Biology at the Scripps Research Institute (Goodsell, 2010) suggests that every living thing on Earth uses a similar set of molecules to eat, to breathe, to move, and to reproduce. There are molecular machines that do the myriad things that distinguish living organisms that are identical in all living cells. This nanoscale machinery of cells uses four basic molecular plans with unique chemical personalities: nucleic acids, proteins, lipids, and polysaccharides.

It is useful to take a deeper look into each of these molecular plans.

Nucleic Acids

Nucleic acids basically encode information. They store and transmit the genome, the hereditary information needed to keep the cell alive. They function as the cell's librarians and contain information on how to make proteins and when to make them.

They are hence, the keepers of a cell's knowledge, its wisdom, its ability to make laws, the vehicle to spread knowledge within cells and to the next generation of cells. Being so, one can see that there is similarity with the set for system-knowledge highlighted earlier in Chapter 2.2. Reproducing Equation 2.2.3:

S_{System_K}
$\ni$ $[Wisdom, Law\ Making, Spread\ of\ Knowledge\ ...]$

Nucleic acids can therefore be thought of as a precipitation of system-knowledge at the cellular level.

Hence, a nucleic acid will have a generalized signature, as in Equation 5.1.1, derived from the system-knowledge family:

$$Sig_{nucleic\ acid} = \frac{Xa +}{\overline{Yb_{0-n}}} \quad where \begin{bmatrix} X \in [S_{System_K}] \\ Y \in [S_{System_{P_r}}, S_{System_P}, S_{System_K}, S_{System_N}] \\ a, b\ are\ integers; a > b \end{bmatrix}$$

Eq 5.1.1: Nucleic Acid

The primary element X could be an attribute or function such as 'keeper of knowledge'. Secondary elements Y could be 'protein laws', 'generational knowledge', amongst others. The collectivity of elements as per the equation would specify the character of a nucleic acid.

DNA and RNA, two types of nucleic acids, would hence have the equations as specified by Equation 5.1.2 and 5.1.3 respectively.

$$Sig_{DNA} = \frac{Xa +}{\overline{Yb_{0-n}}} \quad where \begin{bmatrix} X \in [S_{System_K}] \\ Y \in [S_{System_{P_r}}, S_{System_P}, S_{System_K}, S_{System_N}] \\ a, b\ are\ integers; a > b \end{bmatrix}$$

Eq 5.1.2: DNA

$Sig_{RNA} =$

$$Xa + \overline{Yb_{0-n}} \ where \begin{bmatrix} X \in [S_{System_K}] \\ Y \in [S_{System_{Pr}}, S_{System_P}, S_{System_K}, S_{System_N}] \\ a, b \ are \ integers; a > b \end{bmatrix}$$

Eq 5.1.3: RNA

The primary element X in (5.1.2) and (5.1.3) would be the same as that for nucleic acids (5.1.1). The secondary elements Y however will be a larger and more specific set with many elements in common with (5.1.1).

Proteins

Proteins are the cells work-horses. Look anywhere in a cell and one will see proteins at work. Proteins are built in thousands of shapes and sizes, each performing a different function. As Goodsell describes, "some are built simply to adopt a defined shape, assembling into rods, nets, hollow spheres, and tubes. Some are molecular motors, using energy to rotate, or flex, or crawl. Many are chemical catalysts that perform chemical reactions atom-by-atom, transferring and transforming chemical groups exactly as needed." With their wide potential for diversity, proteins are constructed to perform most of the everyday tasks of the cells. In fact human cells build around 30,000 different kinds of proteins to execute on the diverse array of cellular level tasks.

Proteins hence, exist for service, to bring about perfection at the level of the cell, are characterized by extreme diligence and perseverance, and so on. Being so, one can see that there is similarity with the set for system-presence highlighted earlier in Chapter 2.2. Reproducing Equation 2.2.1:

$$S_{System_{Pr}} \ni [Service, Perfection, Diligence, Perseverance, \ldots]$$

Proteins can therefore be thought of as a precipitation of system-presence at the cellular level.

Hence, a protein could have a generalized signature, as in Equation 5.1.4, derived from the system-presence family:

$$Sig_{protein} = Xa + \overline{Yb_{0-n}} \quad where \begin{bmatrix} X \in [S_{System_{Pr}}] \\ Y \in [S_{System_{Pr}}, S_{System_P}, S_{System_K}, S_{System_N}] \\ a, b \text{ are integers}; a > b \end{bmatrix}$$

Eq 5.1.4: Protein

This could yield a vast number of functional proteins. In fact it may be possible that the 30,000 or so known proteins created by the human cell could each be specified by a signature equation of this nature. It may be possible to map existing proteins to functionality as suggested by the four sets of molecular plans.

Consider Insulin, for example. Insulin regulates the metabolism of carbohydrates, fats and protein by promoting the absorption of, especially, glucose from the blood into fat, liver and skeletal muscle cells. Equation 5.1.5 for Insulin would hence be:

$$Sig_{insulin} = Xa + \overline{Yb_{0-n}} \quad where \begin{bmatrix} X \in [S_{System_{Pr}}] \\ Y \in [S_{System_{Pr}}, S_{System_P}, S_{System_K}, S_{System_N}] \\ a, b \text{ are integers}; a > b \end{bmatrix}$$

Eq 5.1.5: Insulin

The primary element X could be an attribute or function such as 'workhorse. Secondary elements Y could be 'metabolic regulation, 'glucose absorption', 'blood to fat channel', amongst others. The collectivity of elements as per the equation would specify the character of insulin.

Consider Histones as another example. They are the chief protein components of chromatin, acting as spools around which DNA winds, and playing a role in gene regulation. Without histones, the unwound DNA in chromosomes would be very long (a length to width ratio of more than 10 million to 1 in human DNA). Equation 5.1.6 for Histones would hence be:

$$Sig_{histones} = Xa + \overline{Yb_{0-n}} \quad where \begin{bmatrix} X \in [S_{System_{Pr}}] \\ Y \in [S_{System_{Pr}}, S_{System_{P}}, S_{System_{K}}, S_{System_{N}}] \\ a, b \text{ are integers}; a > b \end{bmatrix}$$

Eq 5.1.6: Histones

The secondary element Y would have elements such as 'gene regulation', 'unwound DNA management', amongst others.

Lipids

Lipids by themselves are tiny molecules, but when grouped together form the largest structures of the cell. When placed in water lipid molecules aggregate to form huge waterproof sheets. These sheets easily form boundaries at multiple levels and allow concentrated interactions and work to be performed within a cell. Hence, the nucleus and the mitochondria are contained within lipid-defined compartments. Similarly, each cell itself is contained within a lipid-defined boundary.

Lipids are therefore promoters of relationship, of harmony in the cell, of nurturing the cell-level division of labor, of allowing specialization and uniqueness to emerge, hence perhaps of earlier forms of compassion and love, and so on. The notion of such early forms of compassion is consistent with the biologist's perspective that at some point a gene for compassion was developed in pre-human species (Wright, 2009). Being so, one can see that there is similarity with the set for system-nurturing highlighted earlier in Chapter 2.2. Reproducing Equation 2.2.4:

$$S_{System_N} \ni [Love, Compassion, Harmony, Relationship \; \ldots]$$

This function of harmonization suggests that lipids can therefore be thought of as a precipitation of system-nurturing at the cellular level.

Lipids could have a generalized signature, as in Equation 5.1.7, derived from the system-nurturing family:

$$Sig_{lipid} = \frac{Xa +}{\overline{Yb_{0-n}}} \; where \begin{bmatrix} X \in [S_{System_N}] \\ Y \in [S_{System_{P_T}}, S_{System_P}, S_{System_K}, S_{System_N}] \\ a, b \; are \; integers; a > b \end{bmatrix}$$

Eq 5.1.7: Lipid

Specific lipids such as monoglycerides and phospholipids cold have the following equations:

$$Sig_{monoglyceride} = \frac{Xa +}{\overline{Yb_{0-n}}} \; where \begin{bmatrix} X \in [S_{System_N}] \\ Y \in [S_{System_{P_T}}, S_{System_P}, S_{System_K}, S_{System_N}] \\ a, b \; are \; integers; a > b \end{bmatrix}$$

Eq 5.1.8: Monoglyceride

$$Sig_{phospholipids} = Xa + \overline{Yb_{0-n}} \quad where \begin{bmatrix} X \in [S_{System_N}] \\ Y \in [S_{System_{Pr}}, S_{System_P}, S_{System_K}, S_{System_N}] \\ a, b\ are\ integers; a > b \end{bmatrix}$$

Eq 5.1.9: Phospholipids

The primary element X shared by each of the lipids could be an attribute or function such as 'compartmentalization'. Secondary elements Y could be of the nature of 'work breakdown', 'intra-cell love', amongst others, and would vary with each different kind of lipid.

Polysaccharides

Polysaccharides are long, often branched chains of sugar molecules. Sugars are covered with hydroxyl groups, which associate to form storage containers. As a result polysaccharides function as the storehouse of cell's energy. In addition polysaccharides are also used to build some of the most durable biological structures. The stiff shell of insects, for example are made of long polysaccharides.

Polysaccharides function to create energy, power, courage, strength thereby readying the cell for adventure, and so on. Being so, one can see that there is similarity with the set for system-power highlighted previously in Chapter 2.2. Reproducing Equation 2.2.2:

$$S_{System_P} \ni [Power, Courage, Adventure, Justice, \ldots]$$

Providing energy and strength, polysaccharides can be thought of as a precipitation of system-power at the cellular level.

Polysaccharides could have a generalized signature, as in Equation 5.1.10, derived from the system-power family:

$$Sig_{polysaccharide} = \frac{Xa+}{Yb_{0-n}} \quad where \begin{bmatrix} X \in [S_{System_P}] \\ Y \in [S_{System_{Pr}}, S_{System_P}, S_{System_K}, S_{System_N}] \\ a, b \; are \; integers; a > b \end{bmatrix}$$

Eq 5.1.10: Polysaccharide

Glycogen is an example of a polysaccharide. Glycogen forms an energy reserve that can be quickly mobilized to meet a sudden need for glucose, but one that is less compact and more immediately available as an energy reserve than say triglycerides. Equation 5.1.11 for Glycogen follows:

$$Sig_{glycogen} = \frac{Xa+}{Yb_{0-n}} \quad where \begin{bmatrix} X \in [S_{System_P}] \\ Y \in [S_{System_{Pr}}, S_{System_P}, S_{System_K}, S_{System_N}] \\ a, b \; are \; integers; a > b \end{bmatrix}$$

Eq 5.1.11: Glycogen

Cellulose is another example of a polysaccharide. Cellulose is a polymer made with repeated glucose units bonded together by beta-linkages. Humans and many animals lack an enzyme to break the beta-linkages, so they do not digest cellulose. Equation 5.1.12 for Cellulose follows:

$$Sig_{cellulose} = \frac{Xa+}{Yb_{0-n}} \quad where \begin{bmatrix} X \in [S_{System_P}] \\ Y \in [S_{System_{Pr}}, S_{System_P}, S_{System_K}, S_{System_N}] \\ a, b \; are \; integers; a > b \end{bmatrix}$$

Eq 5.1.12: Cellulose

The primary element for the preceding polysaccharides X would be along the lines of 'energy storage'. The secondary elements may vary, with a Y element for Glycogen being 'rapid energy deployment' for example, and a Y element for Cellulose being 'bonded energy', for example.

Generalized Equations for Cells

Based on the preceding examples a generalized equation for the signature of molecular plans would be as in Equation 5.1.13:

$$Sig = Xa + \overline{Yb_{0-n}} \quad where \begin{bmatrix} X \in [S_{System_{P_r}}, S_{System_P}, S_{System_K}, S_{System_N}] \\ Y \in [S_{System_{P_r}}, S_{System_P}, S_{System_K}, S_{System_N}] \\ a, b \text{ are integers}; a > b \end{bmatrix}$$

Eq 5.1.13: Generalized Equation for Molecular Plan

Using the same logic it is also possible that there could be more nucleic acids, more lipids, more polysaccharides, and even more proteins than are currently known or currently exist.

But further it could be that the genome's reportedly vast tracts of non-coding DNA is reserved for the advances that will inevitably take place as more and more possibilities expressed by such generalized equations come to bear.

Referring to Equation 2.5.6, Generalized Equation of Innovation introduced in Chapter 2.5, this notion of further adaptability implies that the set x_U in the "untransformed" layer U, is likely a 'relatively' as opposed to an absolutely untransformed set, and therefore the starting point for a more sophisticated transformation, which will in turn become the starting point for a further transformation. If so, then the generalized form of the equation, as in Equation 5.1.14, would become:

$$Innovation_{orientation-x} = \left(\begin{bmatrix} M_3 \rightarrow System_X \\ (\uparrow F \rightarrow I) \\ M_2 \rightarrow S_{System_X} \\ (\uparrow Sig \rightarrow F) \\ M_1 \rightarrow Sig_x \\ (\uparrow > P_{x)} \\ U \rightarrow x_U \end{bmatrix} TC \rightarrow x_T \text{, } where \begin{bmatrix} x_U \ni [\ldots] \\ x_T \ni [\ldots] \end{bmatrix} \right)_{\langle x_U | x_T \rangle}$$

Eq 5.1.14: Evolving Form of Generalized Equation of Innovation

Here $\langle x_U | x_T \rangle$ signifies that x_T becomes the new starting point x_U as the process of adaptability continues.

A key difference at the cellular level will likely be in the way the Transformation Circle, TC, functions. The TC was introduced and discussed in Chapter 2.5. Essentially it models a process for the varying rate of transformation with which the dynamics of innovation of any system may proceed.

Hence, for complex behavioral systems whether humans, teams, corporations, markets, or systems, the transformation of the untransformed set, x_U, proceeds in the direction from the untransformed layer, U, through the meta-levels, to M_3. The trigger for this journey is the overcoming of habitual patterns for a specific orientation: $> P_x$. The notion of choice or free-will is therefore inherent. No transformation begins without the organization initiating the trigger conditions. Further, as already discussed, the synthesis or integration is modeled as a function of sin, e, and π.

At the cellular level though it seems that the synthesis between layers happens from M_3, through the meta-levels, to the outer or untransformed layer, U. One can conclude this because there already appears to be a high degree of perfection at the cellular level, and this is put in place automatically, as it were. There does not appear to be free-will or choice as the initiating trigger, in the way that it occurs for a complex organization.

Hence, the Transformation Circle, TC, appears to fulfill the synthesis in another way. Using a system of polar coordinates, since the TC is envisioned as 4 concentric circles, then the radius 'r', measuring from the core to the untransformed layer, indicates the extent of the synthesis. With a shorter 'r' extending only across one or two layers the synthesis is incomplete. When extended across all concentric circles, the synthesis is complete. The time taken to achieve this synthesis is measured by θ. If it is instantaneous, then $\theta = 0$. If it occurred in full-cycle it would be $\theta = 360$, and so on.

Once the synthesis is complete, then the outer layer automatically follows certain rules imposed on it by the meta-layers. This action in organizations, such as the cell, is in stark contrast to the complex organization where the rules need to be changed through deliberation.

Chapter 5.2: Sensations, Urges, Feelings, Thoughts

As human beings we also experience sensations, urges and desires and wills, feelings and emotions, and thought.

But could it be that these are also projections or emergences of the properties of Light, and therefore can be given a mathematical form?

Sensations

Sensations are those things we experience with our senses. We see things, hear things, and smell things, taste things, can touch things. This ability to enter into relationship with objects through sensation is nothing other than a result of the emergence of Light's property of Presence. We become present to Presence through the device of sensation. Sensation can be thought of as the means by which this property of Light – Presence - molds or ingrains itself in us as human beings. Its potentiality, all which is contained in this aspect of Light, becomes available to us through the power of sensation. Hence an equation, Equation 5.2.1, will generally represent the family of sensations. Some elements that it would comprise of may be 'tangible', 'take notice of', amongst others.

$$Sig_{sensation} = Xa + \overline{Yb_{0-n}} \quad where \begin{bmatrix} X \in [S_{System_{Pr}}] \\ Y \in [S_{System_{Pr}}, S_{System_P}, S_{System_K}, S_{System_N}] \\ a, b \text{ are integers}; a > b \end{bmatrix}$$

Eq 5.2.1: Sensation

There could also be equations for hearing, seeing, tasting, touching, and smelling.

But there is also a deeper experience of sensation that is possible. When we see things, for instance, what are we seeing? Is it just the surface rendering of the play of matter, or do we see that the fullness of Light is still there, with all its potentiality and possibility, in the smallest thing we look at? Do we see that the whole universe and more is present in all its fullness in the least thing that we easily ignore, or belittle, or loathe? When we touch things is it the seeming concreteness of the play of the particles or atoms or chains of molecules that we touch? Or is it the Love and Light and the vastness of all that IS that allows itself to be as a small corner that we touch so as to make infinity be felt by something so finite?

Such a deeper contact offered through sensation suggests a subset of (5.2.1) with secondary elements perhaps described as 'fullness of Light', 'contacting infinity', amongst others, thus also yielding an equation form (5.2.2):

$$Sig_{deeper-sensation} = \frac{Xa + }{Yb_{0-n}} \quad where \begin{bmatrix} X \in [S_{System_{Pr}}] \\ Y \in [S_{System_{Pr}}, S_{System_{P}}, S_{System_{K}}, S_{System_{N}}] \\ a, b \; are \; integers; a > b \end{bmatrix}$$

Eq 5.2.2: Deeper Sensation

Urges, Desires & Wills

Urges and desires and wills are similarly a play of the emergence of Light's property of Power. In the mystery of focus, the vastness of Light has projected itself in us into an apparent smallness that is in reality everything that is. And this smallness is trying through urge and desire and will to connect viscerally or even intentionally to other smallnesses that similarly are nothing other than the fullness of Light projected into a small smorgasbord of selected function. So the urge or desire for food, or companionship, or of possession, or of climbing a peak, is nothing other than Light's compressed property of Power, trying to reach more of the fullness that it is through a fulfillment of the urge or desire or will that it masquerades as. Hence urges can be represented as Equation 5.2.3:

$$Sig_{urges} = \frac{Xa + }{Yb_{0-n}} \quad where \begin{bmatrix} X \in [S_{System_P}] \\ Y \in [S_{System_{P_r}}, S_{System_P}, S_{System_K}, S_{System_N}] \\ a, b \ are \ integers; a > b \end{bmatrix}$$

Eq 5.2.3: Urges

Elements may be of the type of 'grasp', 'possess', 'deeply connect', amongst others.

Feelings & Emotions

Feelings and emotions are a play of the emergence of Light's property of Harmony or Nurturing. Its instrument is the Heart and it generates an array of emotions that are an indication or active radar of whether we are moving toward or away from a reality of harmony, whether based on our small self or some larger Self of Light. Gradually, by navigating with these emotions and feelings we can get to a state where we always feel positive emotions which basically means we have more truly entered into relationship with some larger continent of Light. An equation for feelings is as represented by Equation 5.2.4:

$$Sig_{feelings} = \frac{Xa + }{Yb_{0-n}} \quad where \begin{bmatrix} X \in [S_{System_N}] \\ Y \in [S_{System_{P_r}}, S_{System_P}, S_{System_K}, S_{System_N}] \\ a, b \ are \ integers; a > b \end{bmatrix}$$

Eq 5.2.4: Feelings

Thoughts

Thoughts are a play of the emergence of Light's property of Knowledge. Through the thought we can become greater or conceptualize things greater or begun to enter into relationship with some things other than our small self. Thought allows us to connect to more "othernesses" or even the oneness of the reality of Light.

$$Sig_{thoughts} = \frac{Xa +}{Yb_{0-n}} \; where \begin{bmatrix} X \in [S_{System_K}] \\ Y \in [S_{System_{Pr}}, S_{System_P}, S_{System_K}, S_{System_N}] \\ a, b \; are \; integers; a > b \end{bmatrix}$$

Eq 5.2.4: Thoughts

So we can see that the very substance of our becoming – sensations, urges, wills, desires, feelings, emotions, thoughts – are nothing other than an emergence of the four properties of light of presence, power, knowledge, and harmony.

SECTION 6: EMERGENCE OF COMPLEX ORGANIZATIONS

Having traced the emergence of the four properties of Light through the electromagnetic spectrum, through matter, through life, and even through human becoming as manifest in sensation, urges, feelings, and thoughts we now turn our attention to study the phenomenon of emergence in complex organizations.

We begin with deriving an equation for sustainable organization. We then apply this to gain insight into modern-day organization, and suggest the requirements for a sustainable global civilization.

Chapter 6.1: Sustainability of Organizations

We have traced the emergence of the four properties of Light through the electromagnetic spectrum, through matter, through life, and even through human becoming as manifest in sensation, urges, feelings, and thoughts. At each of these levels it seems it is a balanced combination of all four of the properties that creates stability and makes that structure or organism or being sustainable. And this has to be since the four properties occur simultaneously in light and must be impelled to hold to that relationship even when projected from it. Hence it should be possible to deductively arrive at an equation for organizational sustainability given the four-fold emergence of properties of light through subsequent layers of manifestation.

To triangulate, though, we briefly trace the four-fold emergence, reinforce the significance of this and inductively arrive at an equation for sustainability that we would have arrived at deductively anyway.

Starting with the electromagnetic spectrum we see that the spectrum as a whole is defined or architected by the speed of light c, as an emergence of nurturing, the wavelength λ, as an emergence of knowledge, the frequency f, as an emergence of power, and mass-potential as an emergence of presence. These occur simultaneously and constitute the wholeness of the electromagnetic spectrum.

Similarly, at the level of the quantum it can be seen that the integrity and functioning of the atom also depends on the integration of all four properties of light. Hence quarks, an emergence of knowledge, leptons, an emergence of power, bosons, an emergence of harmony or nurturing, and the Higgs-boson, an emergence of presence, act together to create the structure and functionality of every single atom.

What this reinforces is that every single thing must have been created through some integration and balance of the four properties of light.

Going up the organizational scale to the next level of complexity, to the level of the atom, the same pattern is found again: every single atom belongs to one of four groups, and in unison these four groups orchestrate the set of combinations of known compounds.

Hence, Alkali Metals and Alkali Earth Metals configured by the s-Group, from the family of system-power, Metals, Metalloid, Non-Metals, Halogens, and Noble Gases configured by the p-Group, from the family of system-knowledge, Transition Metals configured by the d-Group, from the family of system-presence, and Lanthanides and Actinides configured by the f-Group, from the family of system-nurturing, act together to create the complex array of compounds that form the entire material from which the physical and subsequently even the organic world around us is constructed.

Going up the organizational scale to another level of complexity, to the level of the cell, the same pattern is found again: basically every single cell of every single living creature that has ever been studied by humankind also has a similar balance and integration of these four sets of forces acting together.

Hence, proteins, from the family of presence, nucleic acids, from the family of knowledge, polysaccharides, from the family of power, and lipids, from the family of nurturing, work together to create the balanced functioning of every living cell.

The hypothesis hence, is that even at larger scale, in considering the most innovative organization at the level of teams, corporations, or markets, a similar integration and balance of the four sets of properties or forces may yield the best results. Recent developments in sustainability investment models ranging from Socially Responsible Investing (Logue, 2008), the Global Reporting Initiative (Willis, 2003), the Dow Jones Sustainability Index (Hope & Fowler, 2007), the Principles of Responsible Investing (Harvard-Edu, 2014), all reinforce the concept of investment criteria becoming broader to be based not only on economic, but on environmental, social, and governance factors as well. Further evidence suggests that financial returns on such broad-based investment models continually beat financial returns on regular funds such as the S&P 500, for instance (Openshaw, 2015). In their study of major transitions in evolution Smith and Szathmary (Smith & Szathmary, 1995) chronicle the development of life toward increasing complexity as an application of successful collaboration and even co-evolution by which species evolve by changing together as system pressures increase. This hypothesis can be summarized by the following graph:

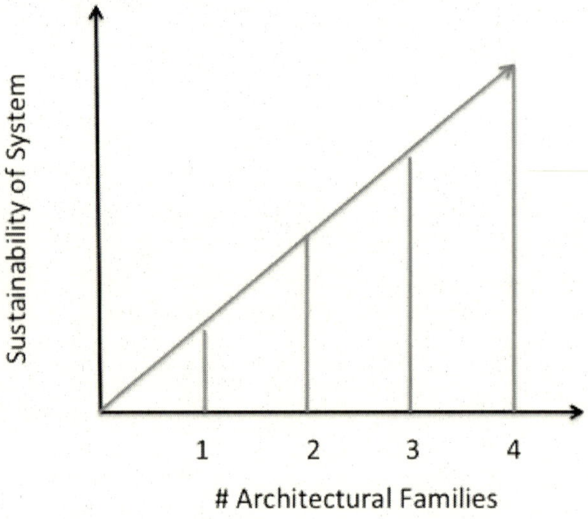

Figure 6.1.1 Sustainability of CAS

An equation for the sustainability of systems, $Sustainability_{Systems}$, where the interaction between the four families of forces is instrumental can be created. Hence, as in Equation 6.1.1:

$$Sustainability_{Systems} \propto Interaction\ (S_{System_{Pr}}, S_{System_{P}}, S_{System_{K}}, S_{System_{N}})$$

Eq 6.1.1: Sustainability of Systems

Chapter 6.2: Creation of a Stable Mega-Organization

As indicated by Equation 6.1.1, a sustainable organization is the result of two dynamics: maturity along each of the architectural force dimensions, and a robust interaction between all for sets of forces. Consider the example of Silicon Valley.

While it may be primarily a commercial organization, yet we begin to glimpse something of what may become possible when the four emergent forces of light can begin to interact with one another in free fashion. That is, without the motive of doing everything for the sake of generating money only. In the case of Silicon Valley we see that the climate and beauty of the Bay Area, thereby likely representing the impetus of nurturing and harmony, began to attract many talented people into the vicinity. Over time the talent pool became progressively diversified. Educational institutes, such as Stanford University and University of Berkeley cropped up and became centers of cutting-edge research, representing the impetus of knowledge. The Armed Forces were attracted to the area for the same reason, and came with their huge requirements for research for research's sake, and their funds in support of the area, representing the impetus of power. Graduates from the universities started companies that began to in turn support the universities with funds. Talent moved around from company to company like people from department to department. Thus we see that even when individual companies failed, Silicon Valley functioned as a larger organization and was able to retain the talent in the area, was able further, to buffer the shocks to some extent, hence preparing the ground for future waves of innovation. This reality of becoming a container for dynamic experimentation perhaps represents the impetus of service in that the container is serving the dynamic components existing within it.

Pragmatically speaking this dynamic of functioning as a larger mega-organization meant that some level of commercial-immunity began to develop, so that people losing their jobs was not necessarily considered as that stressful an event. There was a freeing, thus, from the purely commercial element. If the various powers of beauty, aesthetics, knowledge, power in the form of money, plus the myriad streams of talent – from engineers, scientists, managers, lawyers etc., did not exist in such close proximity, and further, if the various professionals could not thus support each other through their continued informal meetings, through the urge to create the next wave of innovation, through the urge to pursue progress for the sake of progress, then they would have left for other areas and the phenomenon of Silicon Valley would never have been.

Usually modern day organizations cannot provide these various buffers and opportunities for interaction that Silicon Valley provides, and hence when hit with adversity, more often than not, simply crumble. This is perhaps due to the fact that it is always only one-dimension that drives them, and when the health of that is threatened or falls, the organization resorts to tactics that will ensure that that dimension looks good at any cost, in the bargain sacrificing the development of the other dimensions, and therefore its longer term health.

The four-forces led organization will be something like a community. Having attained freedom from an exaggerated commercial impetus, people will 'live' their 'jobs' because it is what fulfills them. Such a freedom is what will allow the four primary powers to manifest to greater and greater degree within them and their environment. Seeing thus, their ability to become centers of knowledge and wisdom, or mutuality and harmony, or power and leadership and energy, or perfection and service, or some unique combination of these primary forces, so increase, a sense of satisfaction with life will more easily accompany all that they continue to do. Under the freer flow of these powers their uniqueness will be refined and flourish, and correspondingly, so too will the uniqueness of their respective organizations.

An organization may be political and hence be primarily driven by the power of leadership and courage, or it may be social and hence be primarily driven by the power of service and perfection, or it may be commercial and hence be primarily driven by the power of mutuality and harmony, or it may be research-oriented and academic and hence be primarily driven by the powers of knowledge and wisdom, but always each of the other powers will also be behind it, fulfilling and completing its primary urge. Thus the dynamic of unique primary powers being supported by an increasing plethora of secondary powers is represented by the following four organizational equations, (6.2.1 – 4) and illustrates how the four powers of light emerge even in complex human-based collectivities:

$$Sig_{Political} = Xa + \overline{Yb_{0-n}} \quad where \begin{bmatrix} X \in [S_{System_P}] \\ Y \in [S_{System_{Pr}}, S_{System_P}, S_{System_K}, S_{System_N}] \\ a, b \text{ are integers}; a > b \end{bmatrix}$$

Eq 6.2.1: Political Organization

$$Sig_{Social} = Xa + \overline{Yb_{0-n}} \quad where \begin{bmatrix} X \in [S_{System_{Pr}}] \\ Y \in [S_{System_{Pr}}, S_{System_P}, S_{System_K}, S_{System_N}] \\ a, b \text{ are integers}; a > b \end{bmatrix}$$

Eq 6.2.2: Social Organization

$$Sig_{Commercial} = Xa + \overline{Yb_{0-n}} \quad where \begin{bmatrix} X \in [S_{System_N}] \\ Y \in [S_{System_{Pr}}, S_{System_P}, S_{System_K}, S_{System_N}] \\ a, b \text{ are integers}; a > b \end{bmatrix}$$

Eq 6.2.3: Commercial Organization

$$Sig_{Academic} = Xa + \overline{Yb_{0-n}} \quad where \begin{bmatrix} X \in [S_{System_K}] \\ Y \in [S_{System_{Pr}}, S_{System_P}, S_{System_K}, S_{System_N}] \\ a, b \ are \ integers; a > b \end{bmatrix}$$

Eq 6.2.4: Academic Organization

Even the number of "unique" people will continue to increase, become guiding lights so that the community will spontaneously begin to move in holistic directions consistent with the urge of people, of the sub-organizations within the community, of the community itself, and of the larger system of which the community or mega-organization is a part.

Even perhaps, something that may be behind the four archetypal powers, something that may be behind the building of all uniqueness – individual and organizational, more of the fullness of light and love, that is, may be compelled to come forward, seeing how developed its means of expression and action have become.

Chapter 6.3: A Sustainable Global Civilization

A brief look at history will also reinforce the idea that it is typically maturity along multiple and distinct dimensions, represented by the four properties of light, and their rich interaction that allows civilizations to endure.

Thus, those civilizations that have endured typically have a balance of all four families (Sri Aurobindo, 1971). Civilizations that have become extinct typically have had a focus on few drivers of innovation. Jared Diamond proposes five interconnected causes of collapse that may reinforce each other: non-sustainable exploitation of resources, climate changes, diminishing support from friendly societies, hostile neighbors, and inappropriate attitudes for change (Diamond, 2005). But these five sources may also be thought of as symptoms that arise due to the failure to adopt the catholicity of the sources of innovation emanating from each of the four sets of families. Further, the historian Toynbee suggested that societies decay because of their over-reliance on structures that helped them solve old problems (Toynbee, 1961). It can be interpreted that being thus biased they are unable to adopt the catholicity of the sources of innovation emanating from each of the four sets of families.

Approaching Civilization from a big-picture, global basis though, the sustainability of humankind will be ensured by a balance of development amongst the four sets of forces. This means that countries must be unique and in such a way that their primary emergence is distributed amongst all four sets of forces. Further, and based on this uniqueness, there must be an open and healthy interaction amongst these centers of uniqueness.

As an illustration, the uniqueness of representative countries in equation form may as follows.

India, for example, in its essence, may be thought of as having an exceptional capacity for penetrating behind the surface, and further of meaningfully synthesizing many streams of development. Its primary power may thus be from the family of knowledge, with a strong secondary driver being its ability to create living harmonies. The equation for India will then likely be represented by Equation 6.3.1:

$$Sig_{India} = \frac{Xa +}{Yb_{0-n}} \text{ where } \begin{bmatrix} X \in [S_{System_K}] \\ Y \in [S_{System_{Pr}}, S_{System_P}, S_{System_K}, S_{System_N}] \\ a, b \text{ are integers}; a > b \end{bmatrix}$$

Eq 6.3.1: India (Knowledge Family)

Japan, in its essence, may be thought of as having a strong and noble warrior nature, along with a refined sense of aesthetics, amongst other qualities. Its equation, Equation 6.3.2 would be of the form:

$$Sig_{Japon} = \frac{Xa +}{Yb_{0-n}} \text{ where } \begin{bmatrix} X \in [S_{System_P}] \\ Y \in [S_{System_{Pr}}, S_{System_P}, S_{System_K}, S_{System_N}] \\ a, b \text{ are integers}; a > b \end{bmatrix}$$

Eq 6.3.2: Japan (Power Family)

UK, in its essence, may be thought of as having a strong ability to create practical, materialistic harmonies, resulting in such things as working parliaments and advanced democracy, for example. Its equation, Equation 6.3.3, would be of the form:

$$Sig_{UK} = \frac{Xa +}{Yb_{0-n}} \text{ where } \begin{bmatrix} X \in [S_{System_N}] \\ Y \in [S_{System_{Pr}}, S_{System_P}, S_{System_K}, S_{System_N}] \\ a, b \text{ are integers}; a > b \end{bmatrix}$$

Eq 6.3.3: UK (Nurturing Family)

Thailand, in its essence, may be characterized by an exceptional sense of hospitality and sweet service, with an attention to detail in the practical arrangement of things. Its equation, Equation 6.3.4, would be of the form:

$$Sig_{Thailand} = Xa + \overline{Yb_{0-n}} \quad where \left[\begin{array}{c} X \in [S_{System_N}] \\ Y \in [S_{System_{Pr}}, S_{System_P}, S_{System_K}, S_{System_N}] \\ a, b \; are \; integers; a > b \end{array} \right]$$

Eq 6.3.4: Thailand (Service Family)

Similarly every nation on earth will have a uniqueness that can be represented by equations belonging to one of the four families.

The second requirement for a sustainable global civilization is a rich and mature interaction between unique nations. But as discussed in Chapter 2.4 on Emergence of Uniqueness, for uniqueness to emerge and mature is a process. The process is represented by Equation 2.4.1 reproduced here for convenience:

$$Sig_E = X \left| \begin{array}{c} C: Sig * mod \left(\int = 1 \right) \\ F: Sig \; mod \; (c) \\ I: Sig \; mod \left(\int \overline{G, e, \pi} \right) \\ M: Sig * mod \; (G) \\ V: Sig * mod \; (e) \\ P: Sig * mod \; (\pi) \end{array} \right|$$

Equations 6.3.1 – 4 represent example of the essence of uniqueness. For these to become living practicalities requires work at many different levels within a nation. If a nation has not really done the work to animate itself with its uniqueness then it may exist in the physical or

P-state, or vital of V-state, or mental or M-state, which by definition lack sufficient maturity, and interaction with other nations is then going to be compromised. Moving to the integral or I-state, will allow a nation to at least not be locked in to a point of view. A consistent I-state practiced by each nation then is the minimum requirement for a sustainable global civilization.

SECTION 7: SUMMARY

This section seeks to tie together the emergence of the properties of light from the form of the electromagnetic spectrum to its current form of civilization.

Chapter 7.1: An Algorithm Tying Together Emergence

In Chapter 5.1 on Cells, the Evolving Form of the Generalized Equation for Innovation (5.1.14) was derived. This equation can be thought of as an algorithm that has allowed emergence to proceed from the mysteries implicit in Light to the verge of sustainable global civilization.

This equation is reproduced here for convenience:

$$Innovation_{orientation-x} = \left(\begin{bmatrix} M_3 \rightarrow System_X \\ (\uparrow F \rightarrow I) \\ M_2 \rightarrow S_{System_X} \\ (\uparrow Sig \rightarrow F) \\ M_1 \rightarrow Sig_x \\ (\uparrow > P_{x)} \\ U \rightarrow x_U \end{bmatrix} TC \rightarrow x_T \text{, } where \begin{bmatrix} x_U \ni [\dots] \\ x_T \ni [\dots] \end{bmatrix} \right)_{\langle x_U | x_T \rangle}$$

The notion of evolution is captured by $\langle x_U | x_T \rangle$, which signifies x_T becoming the new starting point for x_U as the process of adaptability continues.

Hence, if we trace the potential in Light working from the basis of an initial electromagnetic spectrum to the current basis of a possible sustainable global civilization, some of the iterations the equation may have gone through can be envisioned as follows in Equation 7.1.1 – 7, where i < j < k < l < m < n < o. Note that these are only snapshots and always there will be a vast number of iterations as per (5.1.14) that will be required.

$$Emergence_i = \begin{bmatrix} M_3 \rightarrow System_X \\ (\uparrow F \rightarrow I) \\ M_2 \rightarrow S_{System_X} \\ (\uparrow Sig \rightarrow F) \\ M_1 \rightarrow Sig_x \\ (\uparrow > P_{x)} \\ U \rightarrow x_{EM\ Spectrum} \end{bmatrix} TC \rightarrow x_{Quantum\ particles}$$

Eq 7.1.1: EM-Spectrum to Quantum Particles

$$Emergence_j = \begin{bmatrix} M_3 \rightarrow System_X \\ (\uparrow F \rightarrow I) \\ M_2 \rightarrow S_{System_X} \\ (\uparrow Sig \rightarrow F) \\ M_1 \rightarrow Sig_x \\ (\uparrow > P_{x)} \\ U \rightarrow x_{Quantum\ particles} \end{bmatrix} TC \rightarrow x_{Atoms}$$

Eq 7.1.2: Quantum Particles to Atoms

$$Emergence_k = \begin{bmatrix} M_3 \rightarrow System_X \\ (\uparrow F \rightarrow I) \\ M_2 \rightarrow S_{System_X} \\ (\uparrow Sig \rightarrow F) \\ M_1 \rightarrow Sig_x \\ (\uparrow > P_{x)} \\ U \rightarrow x_{Atoms} \end{bmatrix} TC \rightarrow x_{Molecules}$$

Eq 7.1.3: Atoms to Molecules

$$Emergence_l = \begin{bmatrix} M_3 \rightarrow System_X \\ (\uparrow F \rightarrow I) \\ M_2 \rightarrow S_{System_X} \\ (\uparrow Sig \rightarrow F) \\ M_1 \rightarrow Sig_x \\ (\uparrow > P_{x)} \\ U \rightarrow x_{Molecules} \end{bmatrix} TC \rightarrow x_{Cells}$$

Eq 7.1.4: Molecules to Cells

$$Emergence_m = \begin{bmatrix} M_3 \rightarrow System_X \\ (\uparrow F \rightarrow I) \\ M_2 \rightarrow S_{System_X} \\ (\uparrow Sig \rightarrow F) \\ M_1 \rightarrow Sig_x \\ (\uparrow > P_{x)} \\ U \rightarrow x_{Cells} \end{bmatrix} TC \rightarrow x_{Humans}$$

Eq 7.1.5: Cells to Humans

$$Emergence_n = \begin{bmatrix} M_3 \rightarrow System_X \\ (\uparrow F \rightarrow I) \\ M_2 \rightarrow S_{System_X} \\ (\uparrow Sig \rightarrow F) \\ M_1 \rightarrow Sig_x \\ (\uparrow > P_{x)} \\ U \rightarrow x_{Humans} \end{bmatrix} TC \rightarrow x_{Nations}$$

Eq 7.1.6: Humans to Nations

$$Emergence_o = \begin{bmatrix} M_3 \rightarrow System_X \\ (\uparrow F \rightarrow I) \\ M_2 \rightarrow S_{System_X} \\ (\uparrow Sig \rightarrow F) \\ M_1 \rightarrow Sig_x \\ (\uparrow > P_{x)} \\ U \rightarrow x_{Nations} \end{bmatrix} TC \rightarrow x_{Sustainable\ Global\ Civilization}$$

Eq 7.1.7: Nations to Sustainable Global Civilization

Note that in each of these cases the core dynamics are the same and are setup by the seed equation created in the Big Bang, Equation 2.1.6, as derived in Chapter 2.1. The equation is reproduced here for convenience:

$$Light_{Matrix} = \begin{vmatrix} c_\infty: [Pr, Po, K, H] \\ (\downarrow R_{C_K} = f(R_{C_\infty})) \\ c_K: [S_{Pr}, S_{Po}, S_K, S_H] \\ (\downarrow R_{C_N} = f(R_{C_K})) \\ c_{N:} f(S_{Pr}\ x\ S_{Po}\ x\ S_K\ x\ S_H) \\ (\downarrow R_{C_U} = f(R_{C_N})) \\ c_U: [P, V, M, C] \end{vmatrix}$$

Basically the last stable emergence of the properties of Light, and taking (7.1.7) as an example, here it would be 'Nations', is now the new unrefined or U layer. Through a process of overcoming restricting patterns, something of the unique signature of the next stable emergence, in this case 'Sustainable Global Civilization' begins to further manifest and this process continues until it becomes a reality.

REFERENCES

1. Arabatzis, T. 2006. Representing Electrons: A Biological Approach to Theoretical Entities. University of Chicago. Chicago
2. Arkani-Hamed, Nima; Bourjaily, Jacob L.; Cachazo, Freddy; Goncharov, Alexander B.; Postnikov, Alexander; Trnka, Jaroslav (2012). "Scattering Amplitudes and the Positive Grassmannian". arXiv:1212.5605
3. Cottingham, W & Greenwood, D. 2007. An Introduction to the Standard Model of Particle Physics. Cambridge University Press. Cambridge
4. De Broglie, L. 1929. The Wave Nature of the Electron. Nobel Lecture. http://www.nobelprize.org/nobel_prizes/physics/laureates/1929/broglie-lecture.pdf
5. Deep Order Mathematics Videos. 2016. Deep Order Technologies. http://www.deepordertechnologies.com/new-index
6. Diamond, J. 2005. Collapse: How Societies Choose to Fail or Succeed. Viking Books: New York
7. Einstein, A. 1995. Relativity: The Special and General Theory. New York: Broadway Books.
8. Ekspong. 2014. "The Dual Nature of Light as Reflected in the Nobel Archives". *Nobelprize.org.* Nobel Media AB 2014. Web. 15 Oct 2016. http://www.nobelprize.org/nobel_prizes/themes/physics/ekspong/
9. Feinberg, G. 1970. "Particles That Go Faster Than Light". Scientific American. Feb 1970, 69-77.
10. Feynman, RP. 1985. QED The Strange Theory of Light and Matter. New Jersey: Princeton University Press
11. Fuller, B. 1982. Synergetics: Explorations in the Geometry of Thinking. MacMillan Publishing Co.: New York
12. Goodsell, David. 2010. The Machinery of Life. New York: Springer

13. Gray, T. 2009. The Elements: A Visual Exploration of Every Known Atom in the Universe. Black Dog & Levental Publishers. New York.
14. Harvard-Edu. 2014. http://news.harvard.edu/gazette/story/2014/04/harvard-to-sign-on-to-united-nations-supported-principles-for-responsible-investment/
15. Hawking, Stephen. 1988. A Brief History of Time. New York: Bantam Books
16. Heiserman, D. 1991. Exploring Chemical Elements and their Compounds. McGraw-Hill. New York.
17. Hope, Chris; Fowler, Stephen J. (2007). "A Critical Review of Sustainable Business Indices and Their Impact". Journal of Business Ethics. Vol. 76. S. 243–252. Springer: New York
18. Hyperphysics. 2016. Department of Physics and Astronomy. Georgia State University. http://hyperphysics.phy-astr.gsu.edu/hbase/mod3.html#c1
19. Johnson, S. 2010. Where Good Ideas Come From: The Natural History of Innovation. Riverhead Books: New York
20. Kaufmann, S. 1995. At Home in the Universe. New York: Oxford University Press.
21. Laszlo, E. 2014. The Self-Actualizing Cosmos: The Akasha Revolution in Science and Human Consciousness. Inner Traditions: Vermont
22. Logue, A. 2008. Socially Responsible Investing for Dummies. Wiley Publishing: New Jersey
23. Lorentz, H.A. 1925. The Science of Nature. Vol. 25, p 1008. Springer
24. Malik, P. 2009. Connecting Inner Power with Global Change: The Fractal Ladder. New Delhi: Sage Publications
25. NASA-darkmatter. 2016. http://science.nasa.gov/astrophysics/focus-areas/what-is-dark-energy/
26. Ogburn, W. Tomas, D. 1922. Are Inventions Inevitable? A Note on Social Evolution. Political Science Quarterly 37, no.1
27. Olive, K.A et al. 2014. Particle Data Group. Chin. Phys. C, 38, 090001.

28. Openshaw, J. 2015. http://www.marketwatch.com/story/socially-responsible-investing-has-beaten-the-sp-500-for-decades-2015-05-21
29. Pauli, W. 1964. Nobel Lectures, Physics 1942 – 1962. Elsevier Publishing Company. Amsterdam.
30. Pearson, R. 1997. "Frontier Perspectives", Journal of the Center for Frontier Sciences at Temple University. Spring/Summer 1997, Volume 6, Number 2, ISBN: 1062-4767.
31. Perkowitz, S. 2011. Slow Light. London: Imperial College Press
32. Portugali, J., 2012. *Self-organization and the city*. New York: Springer Science & Business Media.
33. Salkind, N. 2007. Encyclopedia of Measurement and Statistics. Thousand Oaks: Sage Publications.
34. Smith, J, Szathmary, E. 1995. The Major Transitions in Evolution. Oxford: Oxford University Press
35. Sri Aurobindo. 1971. Social and Political Thought. Sri Aurobindo Ashram Press: Pondicherry
36. Thomas, D, Brown, J. 2009. Why Virtual Worlds Can Matter. International Journal of Learning and Media 2009 1:1, 37-49
37. Toynbee, A. 1961. A Study of History, Volumes I – XII. Oxford University Press: Oxford
38. Tweed, M. 2003. Essential Elements: Atoms, Quarks, and the Periodic Table. Walker & Copmany. New York.
39. Usó -Doménech, J, Nescolarde-Selva, J, Lloret-Climent, M, Fan, M. 2016. Semiotic open complex systems: Processes and behaviors. Complexity, 21: 388–396. doi:10.1002/cplx.21817
40. Whitaker, A._2006. Einstein, Bohr and the Quantum Dilemma: From Quantum Theory to Quantum Information. Cambridge: Cambridge University Press
41. Wilczek, F. 2016. A Beautiful Question: Finding Nature's Deep Design. New York: Penguin Books
42. Willis, A. 2003. The Role of the Global Reporting Initiative's Sustainability Reporting

Guidelines in the Social Screening of Investment. Journal of Business Ethics. Volume 43. Springer: New York

43. Wright, R. 2009. Evolution of Compassion. https://www.ted.com/talks/robert_wright_the_evolution_of_compassion/transcript?language=en. TED 2009.
44. Yates, F.E. 2012. *Self-organizing systems: The emergence of order.* New York: Springer Science & Business Media.